Mary Lua Adelia Treat

Injurious Insects of the Farm and Garden

Mary Lua Adelia Treat

Injurious Insects of the Farm and Garden

ISBN/EAN: 9783743320741

Manufactured in Europe, USA, Canada, Australia, Japa

Cover: Foto ©berggeist007 / pixelio.de

Manufactured and distributed by brebook publishing software (www.brebook.com)

Mary Lua Adelia Treat

Injurious Insects of the Farm and Garden

OF THE

FARM AND GARDEN.

BY

MARY TREAT.

FULLY ILLUSTRATED.

NEW YORK:
ORANGE JUDD COMPANY,
751 BROADWAY.
1882.

Entered, according to Act of Congress, in the year 1882, by the
ORANGE JUDD COMPANY,
In the Office of the Librarian of Congress, at Washington.

PUBLISHERS' PREFACE.

The assertion that cultivation of all plants, whether on the farm, in the orchard or garden, is largely a struggle with insects, has been strikingly illustrated within the past few years. The standard works upon Entomology include the harmless as well as the injurious insects, and are written with reference to the identification of the species rather than to show how they may be destroyed. In view of the need of a work giving an account of the most destructive insects and the present knowledge of the methods of preventing their ravages, the Publishers invited Mrs. Treat to prepare the present volume. The fact that she has largely availed herself of the works of Prof. Riley, U. S. Entomologist, gives this book a special value.

PREFACE.

Only those who live in the country are aware how much the success of cultivators, whether of farm or garden crops, depends upon insects. There is a surprising lack of knowledge among otherwise well educated people as to the life history of even the most common insects. The questions asked, not only by those in my immediate neighborhood, but by letters from all parts of the country, show how slight is the popular knowledge on this most important branch of Natural History. In view of this, and to bring a knowledge of the most destructive insects within reach of all, this volume has been prepared. It will not be necessary to say to those who make use of this work that the author has availed herself of the permission of Prof. C. V. Riley, to make use of his various contributions to economic entomology.

MARY TREAT.

Vineland, N. J., July, 1882.

CONTENTS.

Introduction 7

INSECTS INJURIOUS TO GARDEN VEGETABLES.

Asparagus 15	Cucumber.
Bean 19	The Pickle-worm 45
Cabbage 21	Melon—The Melon-worm 48
Rape-butterfly 22	Onion 52
Pot-herb Butterfly 24	The Black Onion-fly 52
Southern Cabbage-butterfly . 27	Imported Onion-fly 53
Cabbage-Plusia 29	Parsley and Related Plants 55
Zebra Caterpillar 31	Pea 56
A New Cabbage-worm 33	Radish 61
The Wavy-striped Flea-beetle .. 35	Squash and Pumpkin 61
The Harlequin Cabbage-bug . 37	The Squash-bug 61
Cucumber 42	The 12-spotted Squash-beetle .. 63
Striped Cucumber-beetle ... 42	Tomato 65

INSECTS INJURIOUS TO ROOT CROPS AND INDIAN CORN.

Indian Corn 67	The Potato.
The Corn-worm 68	The Margined Blister-beetle 92
Seed-corn Maggot 72	The Three-lined Leaf-beetle .. 92
The White Grub 73	The Colorado Potato-beetle 94
Cut-worms 78	Sweet Potato 102
Wire-worms 81	Tortoise-beetles 102
" False 82	The Two-striped Sweet-potato Beetle 105
The Potato 83	The Golden Tortoise-beetle. 106
The Stalk-borer 83	The Pale-thighed Tortoise-beetle 108
The Stalk-weevil 85	The Black-legged Tortoise-beetle 109
The Potato-worm 86	Turnip and Ruta-Baga 110
The Striped Blister-beetle . 89	
The Ash-gray Blister-beetle . 90	
The Black-rat and Black Blister-beetles 91	

INSECTS INJURIOUS TO CEREAL GRAINS AND THE GRASS CROPS, INCLUDING CLOVER.

Grains—The Chinch-bug 112	Grains—Northern Army Worm. 130
False Chinch-bugs 117	Wheat-head Army Worm .. 134
The Hessian Fly 120	Clover 135
The Wheat-Midge 123	Clover-seed Midge 135
The Joint-worm 124	Clover Root-borer 136
Army Worms 129	The Clover-worm 137

CONTENTS.

INSECTS INJURIOUS TO FRUIT TREES.

Apple-tree Borer, Round-headed................139
Apple-tree Borer, Flat-headed..144
Apple-twig Borer..............145
Harris' Bark-louse.............147
Oyster-shell Bark-louse........148
Apple-tree Tent-caterpillar.....151
Tent-caterpillar of the Forest..155
Fall Web-worm................160
The Apple-worm — Codling-moth....................161
The Apple-maggot............164
The Apple-curculio............165
The Canker-worm.............166
The Red-humped Caterpillar...170
The Twig-girdler..............171
The New York Weevil.........172
Climbing Cut-worms..........174
The Bag, Basket, or Drop-worm 177
The Slug of Pear and Cherry-tree........................182
The Peach-borer..............183
The Plum-curculio............185
The Periodical, or 17-year Cicada.........................190

INSECTS INJURIOUS TO SMALL FRUITS.

The Currant and Gooseberry...199
   Gooseberry Span-worm....199
   The Imported Currant-worm....................202
   The Native Currant-worm.205
   The Currant Stalk-borer...206
The Strawberry...............206
   The Strawberry-worm.....207
   The Strawberry Leaf-beetle.208
   The Strawberry Leaf-roller.209
   The Strawberry Crown-borer....................209
The Blackberry................210
   Blackberry-borers..........212
The Raspberry................213
   The Snowy Tree-cricket...214
The Grape-vine...............215
   The Hog-caterpillar of the Vine....................215
   The Achemon Sphinx......219
The Grape-vine.
   The Satellite Sphinx........220
   The Abbot Sphinx.........224
   The Blue Caterpillars of the Vine....................226
   The Eight-spotted Forester....................226
   The Beautiful Wood Nymph 228
   The Pearl Wood Nymph...229
   The Grape Leaf-folder.....231
   The Common Yellow Bear.233
   The Grape-vine Plume.....235
   The Grape-berry Moth.....238
   The Grape-vine Flea-beetle.241
   The Spotted Pelidnota.....244
   The Rose-bug, or Rose-chafer..................245
   The Grape Phylloxera.....248
   The Grape Leaf-hopper....259
The Cranberry................260

THE INSECTS OF THE FLOWER GARDEN AND GREENHOUSE.

The Rose-slug.................263
Plant-lice—Aphides............265
Ichnumon Flies on Aphides...265
The Mealy-bug................267

THE ROCKY MOUNTAIN LOCUST.........269

INJURIOUS INSECTS OF THE FARM AND GARDEN.

INTRODUCTION.

It is not the object of this little volume to teach the science of Entomology, or to give the life-history of insects. It is simply intended to group together the most injurious insects with illustrations, that the cultivator may see, at a glance, his enemies, and learn the best known methods of repelling or destroying them. Still there are some points regarding their general structure and changes that may be briefly stated.

The true insects are distinguished from some related animals, the crustaceans, myriapods, and others, by having in their perfect state six legs (the others having either more or none), and generally, though not always, wings.

The insect has three distinct parts: the *head*, in which are the organs of sense; the *thorax*, to which are attached the legs and wings; and the *abdomen*, which contains the reproductive organs. They breathe through breathing holes (*spiracles*) placed along the sides of the body, which communicate with the air tubes within.

Insects exist in four different stages. First, the *egg;* second, the *larva;* third, the *pupa* or *chrysalis;* and fourth, the *imago,* or *perfect insect.*

The parent insect never makes mistakes in providing for posterity, but deposits her eggs on or in just the kind of food her young requires. With most insects the parents live upon a very different kind of food from that on which their numerous offspring feed, and this makes it seem all the more wonderful that they should know so well where to place their eggs. The eggs hatch sometimes within a few days, others take weeks, and some pass the winter months, and hatch with the warmth of the spring sun. It is noticable that those eggs that are not to be hatched until the following spring, are not attached to the leaves or other perishable part of a tree or shrub, but are securely glued to the bark of a twig or branch; they are, moreover, often covered with a kind of varnish which protects them from the rains. Unlike other eggs, those of insects are not injured by intense cold.

The young of all insects, of whatever class, are called *larva* (plural *larvæ*, a Latin word meaning a mask—it being in this stage so unlike the perfect insect that its real form may be said to be masked). Distinct names are popularly given to the larvæ of different insects. The larvæ of Butterflies and Moths are known as *caterpillars;* those of the Beetles are called *grubs*, and when they live in the wood of trees, etc., *borers;* the larvæ of the two-winged flies are known as *maggots*. In a general way, larvæ of most kinds are popularly called "worms," which, though incorrect, has for some insects, as has the term "bug" for others, been adopted by entomologists as the common name for the larvæ of certain species—for example, "Army-worm," "Canker-worm," etc.

The larva is the growing state of the insect, in which it feeds voraciously, moulting, or throwing off its skin from time to time until its full size is attained. The larval stage may last but for a week or two, but in some insects is known to extend over several years. In some insects, as the Mosquitoes and Dragon Flies, the life of the larva is passed entirely in the water.

When the larva has made its full growth it passes into the state of the *pupa*—(the name for an infant rolled up in bandages after the manner of the ancient Romans), this is also called *chrysalis*, from the Greek word for gold, as some have gold-like markings. Most insects are in this state perfectly dormant, while a few, as will be noticed further on, remain active. Some in their last moult appear as if swathed in a hard mummy-like case, others make a cocoon of silken threads, like the Silkworm, in which to assume this state; some make a hollow chamber in the earth for the same purpose; and a number draw together leaves to form a covering to hide them while in the pupa state.

The insect may remain in the pupa state for a few days or weeks, or it may pass the winter in this dormant condition. The methods by which the escape from this imprisonment is made at the proper time, are various and interesting to the observer. In due time it comes forth, and when, as in the case of some moths, it has spread and dried its wings, it seems wonderful that it could have been packed in so small a space.

The perfect insect which is usually provided with wings, is also called the *Imago*, the Latin for an appearance or an image.

In the study of insects, it is convenient to bring them together in what are termed Orders, according to their general resemblances. There are seven of these Orders, each of which is subdivided into families, genera, etc. While entomologists differ as to the minor divisions, these Orders are generally followed in modern works. The first, and regarded as the highest Order is

Order I.—HYMENOPTERA.—The Bees, Wasps, Ants, Ichneumon Flies, etc.

The name *Hymenoptera*, is from the Greek words for "membrane" and "wing." The Greek word *Pteron*, "a wing," plural *Ptera*, "wings," is used in forming the names of all the Orders. The insects of this Order, (with

Fig. 1.—STRAWBERRY FLY.

the exception of the Saw-flies and Horn-tails, which are vegetable feeders), are highly useful to man. They may be regarded as guards over the rest of the insect world, as they serve to keep injurious insects greatly in check. This Order ranks the highest in intelligence, and many of the insects placed here possess wonderful architectural skill. In some of the families the young are provided with nurses, who feed and tend them with the greatest care and apparent affection. Many are provided with stings which are used as weapons of defence.

Order II.—COLEOPTERA.—Beetles, or Shield-winged Insects.

The Greek word *Koleos*, a "sheath," combined with that for "wing," makes up the scientific name of this important Order, which outranks all others in the number and diversity of its species. The insects have two pairs of wings, the upper of which, usually horny or leathery in texture, cover and form a "sheath" for a pair of large membranous

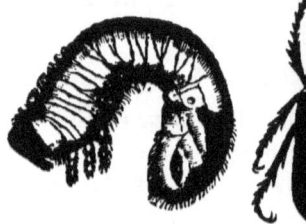
Fig. 2.—LARVA.

Fig. 3.—BEETLE.

wings, which are folded and concealed beneath them. These wing-cases, called *Elytra*, usually meet together with a straight line, or suture, down the back. The larvæ are popularly known as grubs and borers; some live entirely below ground, some are aquatic, while others live upon foliage. This Order includes some of the most injurious insects, and at the same time many carnivorous species, which aid in keeping the vegetable-feeders in check.

Order III.—LEPIDOPTERA.—Butterflies and Moths.

The wings of these insects when touched leave a dust upon the fingers; this, when examined by a magnifier, is

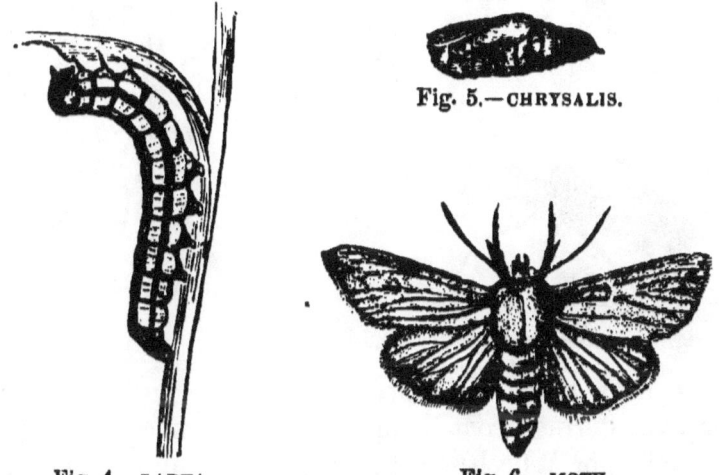

Fig. 5.—CHRYSALIS.

Fig. 4.—LARVA. Fig. 6.—MOTH.

found to consist of minute scales, hence from the Greek word, *Lepis*, a scale, we get the name *Lepidoptera*, or scaly-winged insects. The larvæ, generally known as Caterpillars, present a great variety of forms, and are all strikingly different from the perfect or parent insects. As these are generally vegetable-feeders, the Order is regarded as the most injurious of the seven. Notwithstanding this, in their perfect state, they are among the most beautiful creatures in the insect world. The Order is divided into Butterflies and Moths. The former are day-fliers, and their feelers or *antennæ* are thickened at the end, and terminate in a kind of knob. The moths have their feelers pointed at the tip, and sometimes with small side-branches. They mostly fly at night, but a subdivision of them fly at twilight.

ORDER IV.—HEMIPTERA.—THE TRUE BUGS.

This Order, (the name of which means "half-winged," a portion of the front wings being thick and leathery), includes some very injurious insects—as the Chinch-bug, Squash-bug, Plant-lice, and the disgusting Bed-bug, while some are carnivorous. The

Fig. 7.—WHEEL-BUG (*Reduvius*).   Fig. 8.—HESSIAN FLY.

larvæ have much the appearance of the perfect insect, simply differing from them in the lack of wings. The *Reduvius*, or Wheel-bug, fig. 7, is an example of the carnivorous and useful insects of this Order. The character of the larvæ is seen in the engraving of the Chinch-bug.

ORDER V.—DIPTERA.—TWO-WINGED INSECTS.

This is the only Order of insects that have but **two** wings (a fact expressed in the name). It comprises a great number of species. Not even the *Coleoptera,* can vie with it in numbers. And it embraces some of the most annoying insects—as the Mosquito, Horse-fly, Gnat, and House-fly, also many that are decidedly injurious to vegetation—as the Hessian-fly, Wheat-midge, Onion-maggot, etc., etc. But many of the larvæ act the part of scavengers, and some few are beneficial to the agriculturist—as the *Syrphus* and *Tachina* flies. The young of this Order are known as Maggots.

Order VI.—ORTHOPTERA.—Straight-winged Insects.

The name of this Order is from the Greek, *Orthos,* "straight," the insects have long bodies, straight wings,

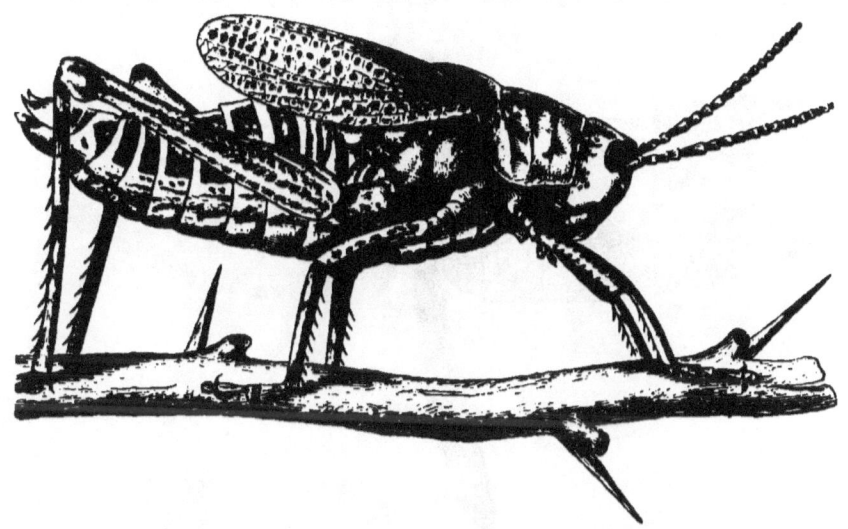

Fig. 9.—THE LUBBER GRASSHOPPER OF FLORIDA.

large heads, and strong jaws. It includes Grasshoppers, Locusts, Cockroaches, Crickets, and Walking-sticks. The larvæ look very much like the parents, except in their usually smaller wings.

Order VII.—NEUROPTERA.—Nerve-winged Insects.

This Order takes in the Dragon-flies, Lace-wings, White-ants, etc. The veins in the wings of these insects are so numerous, that they look like network, and give the name, from *Neuron*, nerve, to the Order. These insects do no harm, with the exception of White-ants, and Book-lice. Some are quite beneficial to man, both in the larval and winged states. The larvæ of the Dragon-flies are aquatic, and exceedingly voracious; they

Fig. 10.—AQUATIC LARVA OF DRAGON-FLY.

Fig. 11.—DRAGON-FLY.

prey upon the larvæ of Mosquitoes, and in the perfect insects destroy vast numbers of winged Mosquitoes.

Insects Injurious to Garden Vegetables.

ASPARAGUS.

THE ASPARAGUS BEETLE.

(*Crioceris asparagi*, Linn.)

About 1860, the Asparagus Beetle was accidentally introduced into Long Island, N. Y., from the other side of the Atlantic; and in a very few years it had increased and multiplied, among the extensive asparagus plantations in that locality, to such an extent as to occasion a dead loss of some fifty thousand dollars in a single county. In the year 1868, it had already crossed over from Long Island on to the adjoining main land; and thence was spread westward.

That our readers may recognize at once this pernicious insect as soon as they see it, we annex figures of it in its various stages. The perfect beetle (shown at *a*, fig. 12, much enlarged, the lines indicating the real size), is of a deep blue-black color, with the thorax brick-red, and some markings of very variable shape and size on the side of its wing-cases. The eggs (*b*) are generally attached to the leaves of the growing asparagus, and are of a blackish color. The larva (magnified at *c*) is of a dull ash color, with a black head and six black legs placed at

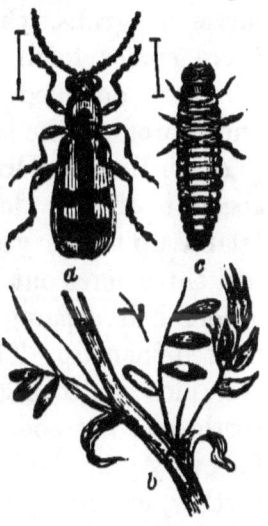

Fig. 12.
ASPARAGUS BEETLE.
(*Crioceris asparagi*.)
a, Beetle; *b*, Eggs; *c*, Larva.

the forward end of the body, the tail end being used as a pro-leg in walking, as with the larvæ of most of the allied beetles. The insect passes the winter under loose bark and in other such sheltered situations, in the perfect or beetle state; and in May, or soon after the season for cutting the asparagus for table use has commenced, it comes forth from its winter quarters and lays the first brood of eggs. These hatch out in about eight days, and by the middle of June the first brood of larvæ are large enough to be noticed, eating the bark off the more tender part of the young stems first, and in default of this consuming the tougher and harder bark of the main stalks. About the end of June they descend to the ground, and either going under the surface of the earth, or hiding under any rubbish that may have accumulated there, form slight cocoons and pass into the pupa state. From these pupæ there bursts forth the same season a second brood of beetles, which lays its eggs as before, and produces about the middle of August a second brood of larvæ or grubs, whence in the same manner as before there comes forth in September the brood of beetles which is destined to pass the winter in the beetle state and reproduce the species in the following spring.

According to Dr. Fitch, who published an excellent account of the depredations of this insect on Long Island up to the year 1862, one asparagus grower there had three acres out of seven "almost ruined;" and four others had asparagus beds so badly injured that they plowed them up. Throughout this entire region the general idea up to 1862 seems to have been, that if this beetle was not soon destroyed, the asparagus would be; for every year the insect appeared to spread further and further, extending already for a distance of at least forty miles along the northern side of Long Island, and every year it got to be more numerous and more destructive. Lime, salt, potash, and a variety of other such applications, had all been tried and found ineffectual as remedies; domestic fowls, which, as Dr. Fitch ascertained, fed greedily upon the beetles, could scarcely be used in sufficient numbers to clear fields of ten and twenty acres in extent; and as to hand-picking twenty-acre fields, especially where the insect is so small, that would be too discouraging an idea to be entertained for a moment.

But in the year 1863, as we learn from Isaac Hicks, of Long Island, a deliverer appeared in the form of a small shining black parasitic fly, probably belonging either to the *Chalcis* or to the *Proctotrupes* Family. Whether this Fly lays its eggs in the eggs of the Asparagus Beetles, or in the larva of that insect, does not seem at present to be clearly ascertained; but if the accounts we have received of it be correct, it must do either one or the other. In the former case, the larva that hatches out from the parasitic egg will consume the egg of the Asparagus Beetle and entirely prevent it from hatching; in the latter case it will destroy the larva before it has time to pass into the perfect state. The result, in either event, will be equally destructive to the bug and beneficial to the gardener. Thus we are told, "although the Asparagus Beetle has not entirely ceased to trouble them upon Long Island since 1863, it yet has never since that year been of any very material damage there. Upon a few farms it still strips the plants in the latter part of summer, but not to much extent or so as to entail any very serious loss."

But the diminution in the numbers of the Asparagus Beetle is probably due in part to artificial, as well as natural causes. The Asparagus growers upon Long Island have introduced a method of fighting the insect, which is founded upon correct principles, and seems to be followed by very gratifying results. Early in the spring, when the Beetle has made its appearance and is ready to lay its eggs, "they destroy," as we are informed, "all the plants upon the farm except the large plants for market, hoeing up all the young seedlings that, as is well known, start from the last year's seed every spring upon the beds." Thus the mother-beetle is forced to lay her eggs upon the large shoots from the old stools; and as these are cut and sent to market every few days, there are no eggs left to hatch out into larvæ for the second brood of beetles.

At first sight we might suppose that it would be possible, by carrying out the above system to its utmost extent, to extirpate the insect entirely. But unfortunately this can not be done. Asparagus, according to Dr. Fitch, has run wild to a considerable extent upon Long Island, "and slender spindling stalks of it may be seen growing in all situations there, by the roadsides, in the fields and in the woods. Thus the Asparagus Beetle has such an abundance of food everywhere presented to it, and the insect is already occupying such an extent of territory, that there seems to be no mode by which it is now possible for us to effect its extermination."

To many persons, perhaps, such a crop as Asparagus may seem of but very trifling importance in a pecuniary point of view. But we have already seen upon how large a scale it is cultivated on Long Island, in the State of New York; and a writer in the "American Journal of Horticulture," who hails from New Jersey, remarks as follows: "We plant Asparagus in great fields of ten to twenty acres. Well planted, it will cost a hundred dollars to set an acre; but it will continue productive for twenty years; and if properly cared for, each acre will clear two hundred dollars annually. There are men all around me who have made small fortunes out of this single article."

BEAN.

THE AMERICAN BEAN-WEEVIL.

(*Bruchus fabæ*, Riley.)

This Weevil appears to be a native American insect and doubtless fed originally on some kind of wild bean (*Phaseolus* or *Lathyrus*), but it was first noticed in our cultivated beans about the year 1861, in Rhode Island, and has since, at different times, suddenly made its appearance in several other parts of the country.

If, as has been supposed, it possibly occurs over large tracts of our country, the fact that, until a few years ago, it had never been collected by any American entomologist, would strongly intimate that, in what may be termed its wild state, it was quite rare and had a limited range. But even if it should occur in this wild state more generally through the country than the facts would lead us to believe, there is nevertheless more danger of its being introduced into a bean field hitherto exempt by the planting of infested cultivated beans, than by its spreading from the wild food. And if once a few buggy beans are planted, they will in a short time infest the other beans cultivated in the neighborhood, so that the man who, year after year, grows his own seed, will suffer as much as the man who originally introduces the weevils from afar.

Except in being smaller, the larva and pupa of this weevil have a close resemblance to those of the Pea-weevil, and its habits are very similar, with the exception that the female deposits a greater number of eggs on a single pod, so that sometimes over a dozen larvæ enter a single bean. As many as fourteen have been counted in one bean, and the space required for each individual to develop is not much more than sufficient to snugly contain the beetle. The little spot where the Pea-weevil entered can always be detected, even in the dry pea, but in the bean these points of entrance become almost entirely obliterated. The cell in which the transformations take place is more perfect and smooth, and the lining is easily distinguished from the meat of the bean by its being more white and opaque. The excrement is yellow, or darker than the meat, and, even where a bean is so badly infested that the inside is entirely reduced to this excrementitious powder, each larva, before transforming, manages to form for itself a complete cell, which separates it from the rest of its brethren. The eye-spot, as in the pea, is perfectly circular and quite transparent in white-skinned varieties, so that infested beans of this kind are easily distinguished by the bluish-black spots which they exhibit (fig. 13, b). Dark beans when infested are not so easily distinguished. The germ is always found either untouched or only partially devoured, even in the worst infested beans, so that when but two or three weevils inhabit a bean, it would doubtless grow; but where the meat is entirely destroyed, as it often is, the bean would hardly grow though the germ remained intact, and it would certainly not produce a vigorous plant. Figure 13, a, gives the weevil magnified, its real size being shown by the small outline at the left.

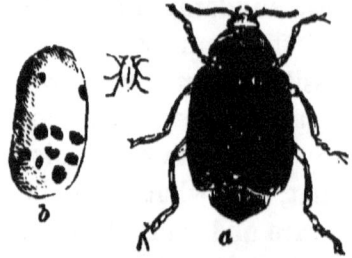

Fig. 13.—THE BEAN-WEEVIL. (*Bruchus fabæ.*) a, Weevil, magnified, the real size in outline; b, Infested Bean.

Some of the beetles are perfected in the fall, but many of them not until the following spring, so that there is the same danger of introducing them in seed-beans as in the case of the Pea-weevil. The remedies and preventives given for the Pea-weevil will of course apply equally well to this, and every bean-grower who reads this should make an effort to keep the scourge out of his own neighborhood, by urging upon others, at the Farmers' Club, or at the meetings of any local societies, the necessity of planting only sound seed, and of thoroughly destroying any that may be received from abroad and found buggy.

CABBAGE.

CABBAGE BUTTERFLIES.

There is a certain group of butterflies, known, scientifically, by the name of *Pieris*, to farmers as "Garden Whites" or "Cabbage Butterflies." They are easily recognized by the following characters: The wings are generally white, with inconspicuous black markings, and occasionally with green or yellow underneath; they are very broad, and have no scallops or indentations in the margin; the hind wings in outline resemble an egg. "The feelers (palpi) are rather slender, but project beyond the head; the antennæ have a short, flattened knob. Their flight is lazy and lumbering. The caterpillars are nearly cylindrical, taper a little towards each end, and are sparingly clothed with short down—which requires a microscope to be distinctly seen. They suspend themselves by the tail and a transverse loop, and their chrysalids are angular at the side and pointed at both ends." (Harris.)

This genus is interesting, though disagreeably so, to every farmer, for the different species are very destructive to various vegetables—among others, cabbages, nasturtium, mignonette, cauliflowers, turnips, and carrots.

We notice only three of the species, as these will serve to indicate the habits of the whole genus—which every farmer should be familiar with, so that he may be able to recognize and destroy such dangerous foes.

THE RAPE BUTTERFLY.

(Pieris rapœ, Schrank.)

This insect has been the occasion of some little speculation and great interest to our New England and Canadian entomologists, inasmuch as it has been introduced to this country from England, and is probably one of the most perfect instances on record of any insect being imported from one country to another and becoming completely naturalized in its new quarters. There does not seem to be the slightest doubt that this is the English species. It was probably introduced in 1856 or 1857. It was first taken in Quebec in 1859, and in 1863 it was captured in large numbers by Mr. Bowles in the vicinity of that city. As the eggs are laid on the under sides of leaves, it was probably introduced in this form, the refuse leaves being thrown out of some ship, after which the larvæ hatched, and, finding themselves in the neighborhood of their food, ate and flourished. Being, moreover, hardy little fellows, they were perfectly able to endure a change of climate. In 1864 it had spread about forty miles from Quebec as a center; in 1866 it was taken in the northern parts of New Hampshire and Vermont; in 1868 it had advanced still further south, and was seen near Lake Winnipesaugee; in 1869 it was taken around Boston, Mass., and a few stray specimens in New Jersey. Since that time it has spread over a wide range of country. The larva and pupa seem to have an unusual power of accommodating themselves to circumstances—for instance, Mr. Curtis, in his "Farm Insects of England," states that the caterpillars have been found feeding on willow.

The larva (fig. 14) is one and one-half inch long; pale-green, finely dotted with black; a yellow stripe down the back, and a row of yellow spots along each side, in a line with the breathing holes. In England and around Quebec it has done immense damage to the cabbages and other *Cruciferæ* (Cress Family), by boring into the very heart of the plant, instead of being content with the less valuable outer portion, as some other species are. On this account the French call it the "*Ver du Cœur*," or Heart-worm. When about to transform, it leaves the plants on

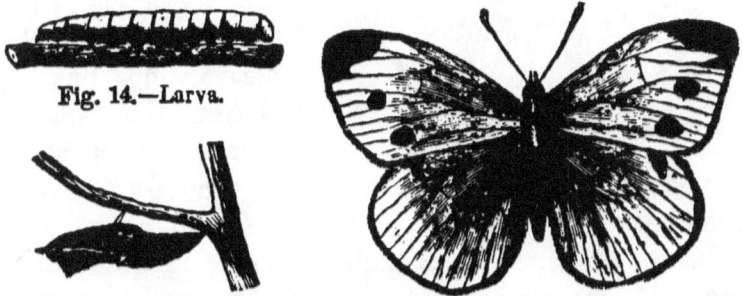

Fig. 14.—Larva.

Fig. 15.—CHRYSALIS. Fig. 16.—RAPE BUTTERFLY (*Pieris rapæ*).—FEMALE.

which it has been living, and fastens itself on the under side of some stone, plank, or fence-rail, where it changes into a chrysalis in the middle or latter part of September, and in this stage it hibernates, producing, in New England, at least, the perfect insect early in April. The chrysalis or pupa (fig. 15) is variable in color, being sometimes yellowish-brown or yellow, and passing thence into green, speckled with minute black dots. The brood of butterflies that emerges from the pupa state in the spring lays eggs shortly afterwards, and these eggs produce caterpillars, which, in their turn, change to chrysalids in June, and in seven or eight days more the butterfly appears, which again lays its eggs for the second brood, which, as before stated, hibernates in the pupa state.

In the perfect butterfly the body and head are black, and the wings white, marked with black, as follows: In the female (fig. 16), a small space at the tip, and three spots on the outer half of the front wings, and one spot on the hind wings; beneath, one spot on the front wings, but none on the hind wings, which are commonly yellowish, sometimes passing into green. The male (fig. 17) has only one spot above and two beneath on the front wings, and a black dash on the anterior edge of the hind wings. There is a variety of the latter sex which has the same markings, but differs from the type in the ground color being canary yellow. Curiously enough, this variety has been taken both in this country and in England.

Fig. 17.
RAPE BUTTERFLY (*Pieris rapæ*).—MALE.

These butterflies occasionally assemble in great numbers. At one time a flight crossed the English channel from France to England, and such was the density and the extent of the living mass, that the sun was completely obscured for a distance of many hundred yards from the people on board a ship that was passing underneath this strange cloud.

THE POT-HERB BUTTERFLY.

(*Pieris oleracea*, Boisd.)

This species has a very wide range, reaching rarely as far south as Pennsylvania, extending eastward to Nova Scotia, and at least as far west as Lake Superior, while in the north it is found as high up as the Great Slave Lake in the Hudson Bay Company's territory. This butterfly (fig. 18, *b*) has a black body; the front wings are white, marked above with black at the base, along the front edge, and at the tip; the hind wings are white above and lemon-yellow beneath, but without markings, except a few black scales at the base.

About the last of May numerous specimens of this species may be seen over cabbage, radish, or turnip beds, or patches of mustard, where, on the under side of the leaves, it deposits its eggs. These are yellowish, nearly pear-shaped, longitudinally ribbed, and one-fifteenth of an inch in diameter, and are seldom laid more than two or three together. In a week or ten days the young caterpillars are hatched; in three weeks more they have attained their full growth, which is an inch and one-half

Fig. 19.—CHRYSALIS.

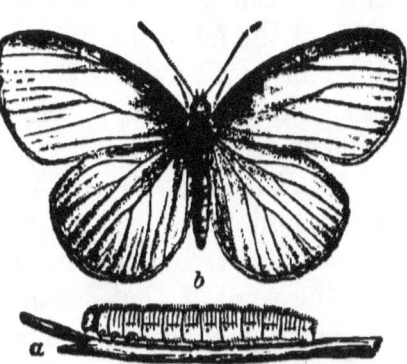

Fig. 18.—POT-HERB BUTTERFLY.
(*Pieris oleracea.*)
a, Larva; *b*, Butterfly.

long. Being slender and green (fig. 18, *a*), they are not readily distinguished from the leaves on which they live. They taper a little toward each end, and are densely covered with hairs. They begin to eat indiscriminately on any part of the leaf. When they have completed the feeding stage, they quit the plants and retire beneath palings, etc., where they spin a little tuft of silk, entangle their hindmost feet in it, and then proceed to form a loop to sustain the front part of the body in a horizontal or vertical position. Bending its head on one side, the caterpillar fastens to the surface, beneath the middle of its body, a silken thread, which it carries across its back and secures on the other side, and repeats this operation until a band, or loop, of sufficient strength is formed. On the next day it casts off the caterpillar skin, and becomes a chrysalis (fig. 19). This is of a pale-green, and sometimes of a white color, regularly and finely dotted with black; the sides of the body are angular, the head is surmounted by a conical tubercle, and over the forepart of the body, corresponding to the thorax of the included butterfly, is a thin projection, having in profile some resemblance to a Roman nose. The insect remains in this stage for ten or twelve days, when the butterfly appears.

In the last of July and first of August, these insects may be seen in large numbers depositing their eggs for a second brood, which, wintering in the pupa state, produces the perfect insect (fig. 18, *b*) the following May.

This butterfly varies considerably. There are never, we believe, perfectly white specimens, though often nearly so. Again, some specimens have very faint indications of spots arranged as in *P. rapæ;* but on the under side are found the widest limits of variation, for not only do the tips of the front wings become distinctly greenish, or lemon-yellow, and the veins of that portion bordered with grayish scales, but the hind wings may also have the ground color distinctly greenish, lemon-yellow, or whitish, and the veins display gray scales on each side.

By taking advantage of the habits of these insects, they might be nearly exterminated. If boards are placed among the infested plants, about two inches above the ground, the caterpillars when about to change will resort to them, and there undergo their metamorphoses. They may then be collected by hand on the under side of the boards, and destroyed. As the butterflies are slow fliers, they may be taken in a net and killed. A short handle, perhaps four feet long, with a wire hoop, and bag-net of muslin or mosquito netting, are all that are required to make this useful implement. The titmouse is said to eat the larvæ, and should therefore be protected and encouraged.

[The descriptions of this and the preceding species are condensed from an article by Chas. S. Minot in "American Entomologist."]

THE SOUTHERN CABBAGE BUTTERFLY.
(Pieris Protodice, Boisd.)

This species, though scarce in the more Northern States, abounds in many of the Southern States, where it takes the place of the two species just described. It often

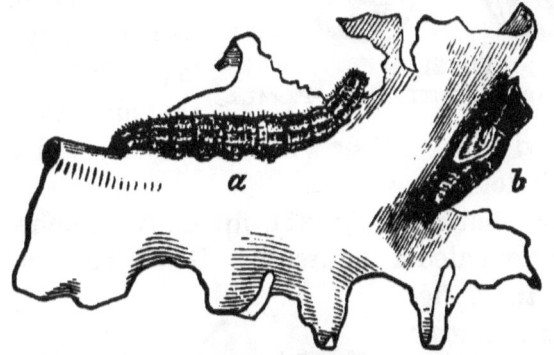

Fig. 20.—SOUTHERN CABBAGE BUTTERFLY (*Pieris Protodice*).
a, Larva; *b*, Chrysalis.

proves exceedingly injurious, and we learn from a Mississippi journal that "there were last year thousands of dollars' worth of cabbages devastated and ruined by worms in the neighborhood of Corinth." We are furthermore told, that cabbages could not, in consequence, be had there even at ten cents per head. The "worm" referred to, was doubtless the species under consideration. It abounds in many parts of Missouri, and especially in the truck gardens around large cities, where it proves quite destructive to the cabbages.

The larva (fig. 20, *a*), may be summarily described as

a soft worm, of a greenish-blue color, with four longitudinal yellow stripes, and covered with black dots. When newly hatched it is of a uniform orange color with a black head, but it becomes dull-brown before the first moult, though the longitudinal stripes and black spots are only visible after said moult has taken place.

Fig. 21.
SOUTHERN CABBAGE BUTTERFLY.—FEMALE.

The chrysalis (fig. 20, *b*.), averages 0.65 inch in length, and is as variable in depth of ground-color, as the larva. The general color is light bluish-gray, more or less intensely speckled with black, with the ridges and prominences edged with buff or with flesh-color, and having larger black dots.

The female butterfly (fig. 21), differs remarkably from the male represented at figure 22. It will be seen, upon comparing these figures that the female is altogether darker than the male. This sexual difference in appearance is purely colorational, however, and there should not be the difference in the form of the wings which the two figures

Fig. 22.
SOUTHERN CABBAGE BUTTERFLY.—MALE.

would indicate, for the hind wings in our male cut are altogether too short and rounded.

This insect may be found in all its different stages through the months of July, August, and September. It hibernates in the chrysalis state. We do not know that it feeds on anything but cabbage, but we once found a male chrysalis fastened to a stalk of the common "Horse Nettle," (*Solanum Carolinense*) which was growing in a cemetery with no cabbages within at least a quarter of a

THE CABBAGE PLUSIA.

(*Plusia brassicæ*, Riley.)

This is the next most common insect which attacks the Cabbage with us, and curiously enough it has never yet

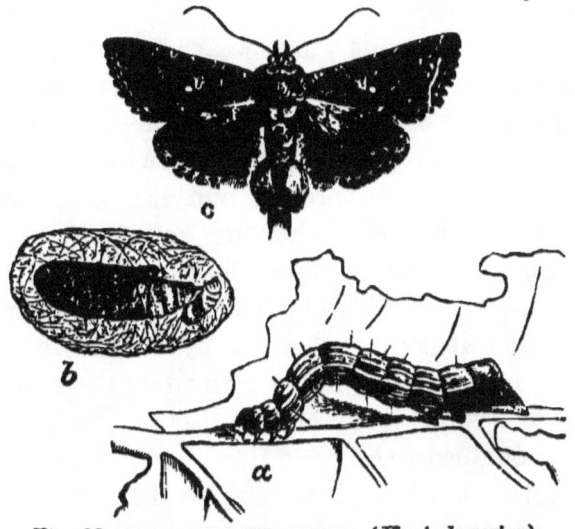

Fig. 23.—THE CABBAGE PLUSIA (*Plusia brassicæ*).
a, Larva; *b*, Chrysalis; *c*, Moth, male.

been described. It is a moth and not a butterfly, and flies by night instead of by day. In the months of August and September the larva (fig. 23, *a*), may be found quite abundant on this plant, gnawing large irregular holes in the leaves. It is a pale-green translucent worm, marked longitudinally with still paler more opaque lines, and like all the known larvæ of the family to which it belongs, it has but two pair of abdominal pro-legs, the two anterior segments, which are usually furnished with such legs in ordinary caterpillars, not having the slightest trace of any. Consequently they have to loop the body in marching, as represented in the figure, and are true Spanworms. Their bodies are very soft and tender, and as they live exposed on the outside of plants, and often rest motionless, with the body arched, for hours at a time, they are espied and devoured by many of their enemies, such as birds, toads, etc. They are also subject to the attacks of at least two parasites and die very often from disease, especially in wet weather, so that they are never likely to increase as rapidly as the butterflies already described.

When full grown, this worm weaves a very thin, loose, white cocoon, sometimes between the leaves on which it fed, but generally chooses some more sheltered situation; and changes to a chrysalis (fig. 23, *b*,) which varies from pale yellowish-green to brown, and has a considerable protuberance at the end of the wing and leg-cases, caused by the long proboscis of the enclosed moth being bent back at that point. This chrysalis is soft, the skin being very thin, and it is furnished at the extremity with an obtuse roughened projection, which emits two converging points, and several short curled bristles, by the aid of which it is enabled to cling to its cocoon.

The moth is of a dark smoky-gray, inclining to brown, variegated with light grayish-brown, and marked in the middle of each front wing with a small oval spot, and a somewhat U-shaped silvery-white mark, as in the engraving. The male (fig. 23, *c*,) is easily distinguished from the female by a large tuft of golden hairs covering a few black ones, which springs from each side of his abdomen towards the tip.

The suggestions given for destroying the larvæ of the Cabbage Butterflies, apply equally well to those of this Cabbage Plusia, and drenchings with a cresylic wash will be found even more effectual, as the worms drop to the ground with the slightest jar.

THE ZEBRA CATERPILLAR.

(*Mamestra picta*, Harr.)

There is another insect which often proves injurious to our cauliflowers and cabbages, though it by no means confines itself to these two vegetables. Early in June the young worms, which are at first almost black, though they soon become pale and green, may be found in dense

Fig. 24.—THE ZEBRA CATERPILLAR (*Mamestra picta*).
a, Larva; *b*, Moth.

clusters on these plants, for they are at that time gregarious. As they grow older, they disperse and are not so easily found, and in about four weeks from the time of hatching, they come to their full growth. Each worm, (fig. 24, *a*,) then measures about two inches in length, and is velvety-black, with a red head, red legs, and with two lateral yellow lines, between which are numerous, transverse white, irregular, zebra-like finer lines, which induced Dr. Melsheimer to call this worm the "Zebra." Though it does not conceal itself, it invariably curls up cut-worm fashion, and rolls to the ground when disturbed.

It changes to the chrysalis within a rude cocoon formed just under the surface of the ground, by interweaving a few grains of sand, or a few particles of whatever soil it happens on, with silken threads. The chrysalis is three-fourths of an inch in length, deep shiny brown, and thickly punctured except on the posterior border of the segments, and especially of those three immediately below the wing-sheaths, where it is reddish and not polished; it terminates in a blunt point ornamented with two thorns. The moth (fig. 24, *b*,) which is called the Painted Mamestra, appears during the latter part of July, and it is a prettily marked species, the front wings of a beautiful and rich purple-brown, blending with a delicate lighter shade of brown in the middle; ordinary spots in the middle of the wing, with a third oval spot more or less distinctly marked behind the round one, are edged and transversed by white lines so as to appear like delicate net-work; a transverse zigzag white line, like a sprawling W, is also more or less visible near the terminal border, on which border there is a series of white specks; a few white atoms are also sprinkled in other places on the wing. The hind wings are white, faintly edged with brown on the upper and outer borders. The head and thorax are of the same color as the front wings, and the body has a more grayish cast.

There are two broods of this insect each year, the second brood appearing in the latitude of St. Louis from the middle of August along into October, and in all probability passing the winter in the chrysalis state, though a few may issue in the fall, and hibernate as moths, or may even hibernate as worms; for Mr. J. H. Parsons, of N. Y., found that some of the worms which were on his Ruta Baga leaves, stood a frost hard enough to freeze potatoes in the hill, without being killed. I have noticed that the spring brood confines itself more especially to young cruciferous plants, such as cabbages, and also on beets, spinach, etc., but have found the fall broods collect in hundreds on the heads and flower-buds of asters, on the Snow-berry or White-berry (*Symphoricarpus racemosus*); on different kinds of Honey-suckle, Mignonette, and on Asparagus; they are also said to occur on the flowers of Clover, and are quite partial to the common Lamb's-quarter, or Goosefoot (*Chenopodium album*). On account of their gregarious habit when young, they are very easily destroyed at this stage of their growth.

A NEW CABBAGE WORM.

(*Pionea rimosalis*, Guen).

Prof. Cyrus Thomas, Carbondale, wrote to the "American Entomologist," in substance, as follows: "I have something new. It is a new Cabbage worm, the larva of *Pionea* [*Orobena*] *rimosalis*, Guen., which appeared late the past season, remaining on the cabbages till toward the end of November. It is very destructive, doing as much injury to my cabbages after it appeared as the imported Cabbage worm (*Pieris rapæ*) which has been very destructive here this season. The larva, when full grown, is six or seven-tenths of an inch long (a 16-legged Pyralid larva); slender, slightly flattened; head shining greenish-yellow; dorsal portion of the body down to the breathing pores purplish-brown; this portion marked with numerous transverse whitish lines, two or three to a segment; a narrow, pale yellow line along the region of the stigmata; underside pale green. In the breeding cages they went down to the soil, but not into it, to pupate; forming a slight, regularly shaped, oval cocoon, thickly covered over with sand.

"Miss Middleton's record shows as follows; "Went into the pupa state September 12th, 13th, and 14th; moths appeared 16th to 22nd, and on to Oct. 1st.

"After this there was another brood of worms, my description having been taken from living specimens, Nov. 21st. The eggs I have not seen, but from the fact that the young feed somewhat together (though not really in companies) I presume a number are laid together. These worms eat, as a general thing, elongate oval holes in the leaves, gradually extending them until nothing but the larger veins remain.

"They also bore directly into the heads, to the depth of, or rather through three or four leaves; a habit, so far as my experience goes, wrongly ascribed to the larva of *P. rapæ*, which will seldom eat through even one leaf of a solid head until it is at least slightly loosened.

"Lime, ashes, brine, salt, elder decoction, and lye as strong as the cabbages can bear, and other substances tried, have even less effect upon them than on the imported Cabbage worm. The lye, fresh made, of strong ashes, did more than anything else tried.

"I have ascertained that some varieties of the cabbage suffer much less from *P. rapæ* than others, and that bringing them forward two or three weeks earlier than usual so as to have the heads pretty well formed before the full brood appears, is also an excellent plan to counteract them."

The editor adds: This is the first instance which has come to our knowledge, of *Pionea rimosalis* injuring cabbage. It is interesting, as illustrating the unity of habit in the genus which essentially feeds on *Cruciferæ*. The larva of *P. forficalis*, L., is very destructive to cabbages in Europe, working very much as Prof. Thomas has described in the case of *P. rimosalis*.

REMEDY FOR CABBAGE WORMS.—Of all the many topical remedies that have been tried for the Imported

Cabbage worm since it first began to spread over the country and to play havoc with our cabbage fields, few, if any, have given entire satisfaction. It is safe to say that the most satisfactory remedy so far discovered is in the use of Pyrethrum. Prof. Riley was the first to apply this in 1879, but did not care to recommend it until further experiments had been made. He has made these since, and caused others to be made by a number of his agents and correspondents. The general experience is most favorable, and he unhesitatingly recommends it for all the different worms affecting the leaves of our cabbage plants. Some have found hot water very effective on a large scale. Living plants will bear, without injury, for a few seconds, water hot enough to kill soft-bodied insects. The water should be at the temperature of about 160° when it reaches the plant; it will cool somewhat during the application, and allowance should be made for this.

THE WAVY-STRIPED FLEA-BEETLE.

(*Haltica* [*Phyllotreta*] *striolata*, Illiger.)

"The Striped Turnip-beetle (fig. 25, *a*,) is less than one-tenth of an inch in length. Its general appearance is black, with a broad wavy yellowish, or buff-colored stripe, on each wing-cover. The larva (fig. 25, *b*,) is white, with a faint darkened or dusky median line on the anterior half of the body, being probably the contents of the alimentary canal seen through the semi-translucent skin. The head is horny and light brown. On the posterior extremity is a brown spot equal to the head in size; and there are six true legs and one

Fig. 25.—WAVY-STRIPED FLEA-BEETLE.
(*Haltica striolata*.)
a, Beetle; *b*, Larva; *c*, Pupa.

proleg. In its form and general appearance it somewhat resembles the larva of the Cucumber-beetle, but it is much smaller. Its motion is slow, arching up the abdomen slightly, on paper or any smooth surface, in such a position that its motions are necessarily awkward and unnatural, because in a state of nature it never crawls over the surface, but digs and burrows among the roots in the ground. Its length is 0.35 of an inch, and breadth 0.06 of an inch. It feeds upon roots beneath the ground.

"The pupa (fig. 25, *c*,) is naked, white, and transforms in a little earthen cocoon, pressed and prepared by the larva, in the ground near its feeding place. This period is short.

"Every gardener knows that these insects are very injurious to young cabbages and turnips as soon as they appear above the ground, by eating off the seed-leaves; he also almost universally imagines that when the second, or tree-plant leaves appear, that the young plant is safe from their depredations; then the stem is so hard that the insect will not bite it, and the leaves grow out so rapidly as not usually to be injured by them. But if we would gain much true knowledge of what is going on around us, even among these most simple and common things, we must learn to observe more closely than most men do.

"The gardener sees his young cabbage plants growing well for a time, but at length they become pale or sickly, wither and die in some dry period that usually occurs about that time, and attributes their death to the dry weather; but if he will take the pains to examine the roots of the plants, he will find them eaten away by some insect, and by searching closely about the roots will find the larva, grub, worm, or whatever else he may choose to call it; from this he can breed the Striped Turnip-beetle, as I have often done.

"I have observed the depredations of these larvæ for ten years, and most of that time had a convincing knowledge of their origin, but only proved it in 1865; since that time I have made yearly verifications of this fact.

"Every year the cabbage plants and turnips in this region receive great damage from these larvæ, and often when we have dry weather, in the latter part of May and early in June, the cabbage plants are ruined. A large proportion of them are killed outright in June, and the balance rendered scarcely fit for planting; but when the ground is wet to the surface all the time by frequent rains, the young plant is able to defend itself much more effectually, by throwing out roots at the surface of the ground, when the main or center root is devoured by the larva; but in dry weather these surface roots find no nourishment and the plant must perish.

"This year I saw these beetles most numerous in early spring, but have often seen them in August and September, so abundant on cabbages, that the leaves were eaten full of holes, and all speckled from their presence, hundreds often being on a leaf; and at this time the entire turnip crop is sometimes destroyed by them, and seldom a year passes without their doing great injury. * * * As the Cucumber-beetle raises its young on the roots of the Gourd Family exclusively, I am led to believe that the Striped Turnip-beetle raises its young always on the roots of the Mustard Family."—[Dr. Henry Shimer, in "American Naturalist," December, 1868.]

THE HARLEQUIN CABBAGE-BUG.

(*Strachia histrionica*, Hahn.)

Cabbage-growers in the North are apt to think, that the plant which they cultivate is about as badly infested by insects as it is possible for any crop to be, without being utterly exterminated. No sooner are the young cabbages above ground in the seed-bed, than they are often attacked by several species of Flea-beetles. By these jumping little pests the seed-leaves are frequently riddled so full of holes that the life of the plant is destroyed; and they do not confine themselves to the seed-leaves, but prey to a considerable extent also upon the young rough leaves. After the plants are set out, the larva of the very same insect is found upon the roots, in the form of a tiny elongate six-legged worm. Through the operations of this subterranean foe, the young cabbages, especially in hot dry weather, often wither away and die; and even if they escape this infliction, there is a whole host of cutworms ready to destroy them with a few snaps of their powerful jaws; and the common White Grub, as we know by experience, will often do the very same thing. Suppose the unfortunate vegetable escapes all these dangers of the earlier period of its existence. At a more advanced stage in its life, the stem is burrowed into by the maggot of the Cabbage Fly (*Anthomyia brassicæ*)—the sap is pumped out of the leaves in streams by myriads of minute Plant-lice covered with a whitish dust (*Aphis brassicæ*)—and the leaves themselves are riddled full of holes by the tiny larva of the Cabbage Tinea (*Plutella cruciferarum*), or devoured bodily by the large fleshy larvæ of several different Owlet-moths.

Severe as are these inflictions upon the Northern cabbage-grower, there is an insect found in the Southern States that appears to be, if possible, still worse. This is the Harlequin Cabbage-bug (*Strachia histrionica*, Hahn, fig. 26, *d*, which is enlarged, the line showing the real size), so called from the gay theatrical Harlequin-like manner in which the black and yellow colors are arranged upon its body. The first account of the operations of this very pretty but unfortunately very mischievous bug appeared in the year 1866 from the able pen of the late Dr. Gideon Lincecum, of Washington county, Texas, and were printed in the "Practical Entomologist" (vol. I, p. 110). His remarks are to the following effect:

"The year before last they got into my garden, and utterly destroyed my cabbage, radishes, mustard, seed turnips, and every other cruciform plant. Last year I did not set any of that order of plants in my garden. But the present year, thinking they had probably left the premises, I planted my garden with radishes, mustard, and a variety of cabbages. By the first of April the mustard and radishes were large enough for use, and I discovered that the insect had commenced on them. I began picking them off by hand and trampling them under foot. By that means I have preserved my four hundred and thirty-four cabbages, but I have visited every one of them daily now for four months, finding on them from thirty-five to sixty full-grown insects every day, some coupled and some in the act of depositing their eggs. Although many have been hatched in my garden the present season, I have suffered none to come to maturity; and the daily supplies of grown insects that I have been blessed with, are immigrants from some other garden.

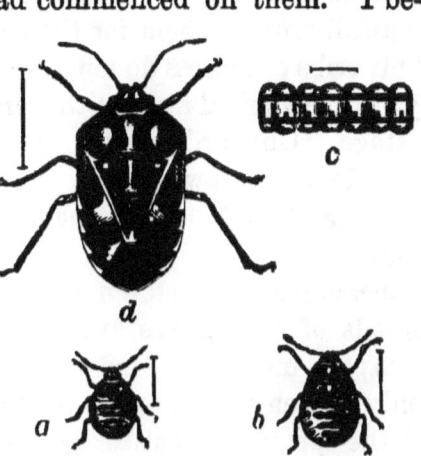

Fig. 26.—HARLEQUIN CABBAGE-BEETLE
(*Strachia histrionica*).
a, Larva; *b*, Pupa; *c*, Eggs; *d*, Beetle.

"The perfect insect lives through the winter, and is ready to deposit its eggs (fig. 26,*c*,) as early as the fifteenth of March, and sooner, if it finds any cruciform plant large enough. They set their eggs on end in two rows, cemented together, mostly on the under side of the leaf, and generally from eleven to twelve in number. In about six days in April—four days in July—there hatches out from these eggs a brood of larvæ (fig. 26, a,) resembling the perfect insect, but has no wings. This brood immediately begins the work of destruction by piercing and sucking the life-sap from the leaves; and in twelve days they have matured. They are timid, and will run off and hide behind the first leaf-stem, or any part of the plant that will answer the purpose. The leaf that they puncture immediately wilts. Half a dozen grown insects will kill a cabbage in a day. They continue through the summer, and sufficient perfect insects survive the winter to insure a full crop of them for the coming season.

"This tribe of insects do not seem liable to the attacks of any of the cannibal races, either in the egg state or any other stage. Our birds pay no attention to them, neither will the domestic fowls touch them. I have, as yet, found no way to get clear of them, but to pick them off by hand."

It appears from this statement that there are at least two broods of the species every year, the first hatching out in April and the second in July; and as it is said that only sixteen or eighteen days elapse from the deposition of the egg to the mature development of the perfect bug, it is not improbable that the species is in reality many-brooded. The eggs, of which we have specimens now before us, are about 0.03 inch in diameter, barrel-shaped, and of a greenish-white color with two broad black bands encircling the staves of the barrel so as to look exactly like hoops. To afford a passage to the young larva, one of the heads of the barrel—the one, of course, that is not glued to the surface of the leaf—is detached by the beak of the little stranger as neatly and as smoothly as if a skillful cooper had been at work on it with his hammer and driver. And yet, instead of employing years in acquiring the necessary skill, the mechanic that performs this delicate operation with unerring precision, is actually not yet born into this sublunary world!

Hitherto it had been generally supposed by entomologists that the Harlequin Cabbage-bug was confined to the most southerly of the Southern States, such as Texas and Louisiana; and it has consequently been called by some "the Texan Cabbage-bug," instead of translating the scientific name and calling it, as we have done, "the Harlequin Cabbage-bug." In September, 1867, however, we received numerous living specimens from Dr. Summerer, of Salisbury, in North Carolina; and from his account it seems to be as great a pest in the gardens of that State as Dr. Lincecum describes it to be in Texas. Hence the species is most probably to be met with, in particular localities and in particular seasons, throughout the Southern States, at least as far north as Tennessee and Arkansas; and we should not be surprised if a few specimens were eventually to turn up in Southern Illinois, and in Southern Missouri.

It is said that no criminal among the human race is so vile and depraved, that not one single redeeming feature can be discovered in his character. It is just so with this insect. Unlike the great majority of the extensive group (*Scutellera* Family, Order of Half-winged Bugs) to which it belongs, it has no unsavory bed-buggy smell, but on the contrary exhales a faint odor which is rather pleasant than otherwise. We have already referred to the beauty of its coloring. As offsets, therefore, to its greediness and its thievery, we have, first, the fact of its being agreeable to the nose, and secondly the fact of its being agreeable to the eye. Are there not certain demons in the garb of angels, occasionally to be met with among the human species, in favor of whom no stronger arguments than the above can possibly be urged?

CUCUMBER.

THE STRIPED CUCUMBER-BEETLE.

(*Diabrotica vittata*, Fabr.)

The Striped Cucumber-Beetle is an insect which annually destroys thousand of dollars' worth of vines in the United States, and for which remedies innumerable—some sensible, but the greater portion not worth the paper on which they are printed—are published every year in some of the agricultural papers.

As everything pertaining to such a very common and destructive insect cannot be too often repeated, I will here relate its habits in the briefest manner.

The parent beetles (fig. 27) make their appearance quite early in the season, when they immediately commence their work of destruction. They frequently penetrate through the cracks that are made by the swelling and sprouting of the seeds of melons, cucumbers, or squashes, and by nipping off the young sprouts, destroy the plant before it is even out of the ground.

Fig. 27.—STRIPED CUCUMBER-BEETLE.

Their subsequent work, when the vines have once pushed forth their leaves, is too well known to need description. Yet notwithstanding the great numbers and the persistency of these beetles, we finally succeed, with the proper perseverance and vigilance in nursing and protecting our vines, until we think they are large enough to withstand all attacks. Besides, by this time, the beetles actually begin to diminish in numbers, and we congratulate ourselves on our success. But lo! All of a sudden, many of our vines commence to wilt, and they finally die outright. No wound or injury is to be found on the vine above ground, and we are led to examine the roots. Here we discover the true cause of death, for the roots are found to be pierced here and there with small holes, and excoriated to such an extent that they present a corroded appearance. Upon a closer examination the authors of this mischief are easily detected, either imbedded in the root, or lurking in some of the corroded furrows. They are little whitish worms, rather more than a third of an inch long, and as thick as a good-sized pin; the head is blackish-brown and horny, and there is a plate of the same color and consistency on the last segment. These worms are in fact the young of the same Striped-Bug which had been so troublesome on the leaves earlier in the season; and that the insect may be as well known in this its masked form, as it is in the beetle state, I present the annexed highly-magnified figures of the worm (fig. 28) showing a back, and fig. 29 a side view. The beetles, while feasting themselves on the tender leaves of the vine, were also pairing, and these worms were hatched from the eggs deposited near the roots by the females. When the worms have become full-grown, which is in about a month after they hatch, they forsake the roots and retire into the adjoining earth, where each one, by continually turning around and around, and compacting the earth on all sides, forms for itself a little cavity, and in a few days throws off its larva skin and becomes a pupa. This pupa is much shorter than was the worm, and the insect lasts in this state about

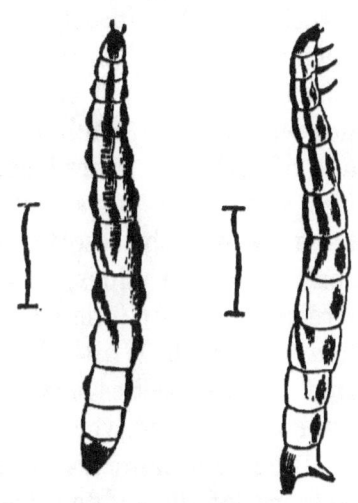

Fig. 28. LARVA. BACK VIEW.   Fig. 29. LARVA. SIDE VIEW.

two weeks, at the end of which time the skin is again moulted, and the perfect beetle form assumed. All the parts of this newly-developed beetle are at first soft, but after remaining motionless in its cell until these soft parts have acquired solidity and strength, it breaks through the walls of its prison and works itself up to the light of day. There are from two to three generations each year, the number varying according to the latitude, or the length of the winter.

Of all the multifarious remedies proposed against the attacks of this insect, there is none so effectual or so cheap in the end, as inclosing the young vines in boxes, which are opened at the bottom, and covered with millinet on the top. Such boxes are made at a trivial cost, and if properly stored away each season after use, will last for many years. Whenever other remedies must from necessity be resorted to, there is nothing better than sprinkling the vines, early in the morning, with Paris Green and Flour (one part of the Green to four or five of flour), or with White Hellebore. It of course follows, that if the beetles are effectually kept off, there will afterwards be no worms at the roots.

Much complaint has been made in various parts of the country, of the sudden death of cucumber and other cucurbitaceous vines, from some unknown cause, and Henry Ward Beecher seems to have suffered in this manner, like the rest of us, but could find no worms in the roots of his vines. I know from experience that such vines are subject to a species of rot in the root, a rot not caused by insects, and for that reason the more serious, since we cannot tell how to prevent it. I have seen whole melon patches destroyed by this rotting of the roots, but in the great majority of instances where I have examined vines that had died from "some unknown cause," I have had no difficulty in either finding the worms of the "Striped Bug" yet at work on the roots, else the unmistakable marks of their having been there. Indeed, by the time a vine dies from the effects of their gnawings and burrowings, the worms have generally become fully grown, and have hidden themselves in their little pupal cavities.

So much for the two borers which have heretofore been known to attack plants belonging to the Gourd family. We have seen how they both bore into the roots of these plants, and how one of them in the perfect state attacks the leaves. No other borers have been known to attack these plants, though the 12-Spotted Diabrotica (*D. 12-punctata*, fig. 30), may be found embedded in the rind of both melons, cucumbers, and squashes. But we now come to a third insect which attacks plants of this same Gourd family. It neither bores into the root, nor devours the foliage, however, but seems to confine itself to the fruit; and I have called it the Pickle Worm, from the fact of its often being found in cucumbers that have been pickled.

Fig. 30.—12-SPOTTED DIABROTICA.

THE PICKLE WORM.

(*Phacellura nitidalis*, Cramer.)

At figure 31 is represented one of these worms, of the natural size. They vary much in appearance, some being of a yellowish-white, and very much resembling the inside of an unripe melon, while others are tinged more or less with green. They are all quite soft and translucent, and there is a transverse row of eight shiny, slightly elevated spots on each of the segments. Along the back and towards the head these spots are larger than at the sides, and each spot gives rise to a fine hair. The specimen from which I obtained my first moth was very light-colored, and these spots were so nearly the color of the body as to be scarcely visible. The head was honey-yellow, bordered with a brown line, and with three black confluent spots at the palpi.

The worms commence to appear, in the latitude of St. Louis, about the middle of July, and they continue their destructive work till the end of September. They bore cylindrical holes into the fruit, and feed on its fleshy parts. They are gross feeders, and produce a large amount of soft excrement. I have found as many as four in a medium-sized cucumber, and a single worm will often cause the fruit to rot. They develop very rapidly, and come to their growth in from three to four weeks. When about to transform, they forsake the fruit in

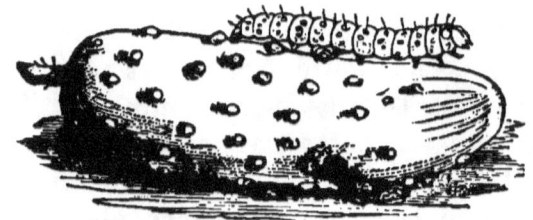

Fig. 31.—PICKLE WORM (*Phacellura nitidalis*).—Larva.

which they had burrowed, and drawing together portions of some leaf that lies on or near the ground, spin a light cocoon of white silk. Within this cocoon they soon become slender brown chrysalids, with the head parts prolonged, and with a very long ventral sheath, which encloses the legs. If it is not too late in the season, the moths issue in from eight to ten days afterwards. The late individuals, however, pass the winter within their cocoons, though, from the fact that some moths come out as late as November, I infer that they may also winter over in the moth state.

The moth produced by this worm (of which figure 32 represents the male) is very strikingly marked. It is of a yellowish-brown color, with an iris-purple reflection, the front wings having an irregular, semi-transparent, dull, golden-yellow spot, not reaching their front edge, and constricted at their lower edge, and the hind wings having their inner two-thirds of this same semi-transparent yellow. The under surfaces have a more decided pearly lustre. The thighs, the breast, and the abdomen below, are all of a beautiful silvery-white, and the other joints of the long legs are of the same tawny or golden-yellow as the semi-transparent parts of the wings. The abdomen of the female terminates in a small, flattened black brush, squarely trimmed, and the segment directly preceding this brush is of a rust-brown color above. The corresponding segment in the male is, on the contrary, whitish anteriorly, and of the same color as the rest of the body posteriorly; and he is, moreover, at once distinguished from the female by the immense brush at his tail, which is generally much larger than represented in the above figure, and is composed of narrow, lengthened (ligulate) scales, which remind one of the petals of the common English Daisy, some of these scales being whitish, some orange, and others brown. This moth was described nearly a century ago by Cramer.

Fig. 32.—MOTH OF PICKLE WORM.

The genus to which it belongs is characterized chiefly by the partly transparent wings, and by the immense scaly brush of the males. The antennæ are long, fine, and thread-like, those of the male being very finely ciliated; the abdomen extends beyond the wings, and the legs are very long and slender. The species are for the most part exotic, and the larvæ of all of them, so far as known, feed on cucurbitaceous plants.

But our Pickle-worm is an indigenous species, and has, doubtless, existed in some part or other of the country, from time immemorial; and now that its habits are recorded, and its history made known, I should not be at all surprised to learn that individuals have suffered from it in years gone by. The French Entomologist, Guenée, gives as its food-plant, a species of potato, and it is just possible that it may not always have fed upon the same plants upon which it was first found in this country. At all events, let us hope that it will disappear as suddenly as it appeared; but should it occur again in great numbers, the foregoing account will enable those who grow melons, cucumbers, or squashes, to understand their enemy, and to nip the evil in the bud, by carefully overhauling their vines in the summer, and destroying the first worms that appear, either by feeding the infested fruit to hogs or cattle, or by killing the worms on the spot.

THE MELON-WORM.

(*Phacellura hyalinatalis*, Linn.)

The Melon-worm is described by Prof. J. H. Comstock, in the Report of the Department of Agriculture, for 1879, as eating cavities into melons, cucumbers, and pumpkins at all stages of growth, and also devouring their leaves. The perfect insect (*Phacellura hyalinatalis*, also written *Phakellura*), has long been well known to entomological collectors from its beauty and abundance in certain localities; but has received almost no attention from economic workers. Guenée in giving its geographical distribution, says: "Very common in all America. I have received it from Brazil, from Columbia, from Hayti, from North America, and from French Guiana."

As to the food plant of the larva Guenée simply states that it lives upon the pumpkins, watermelons, and other cucurbitaceous plants.

In the July, 1875, number of "Field and Forest" a short

account is given of the destruction of a large crop of cucumbers at Indian River, Florida, by these worms. It was stated that they first attacked the bud, then worked into the plant, and eventually killed them out, root and branch. The melon crop in parts of Georgia has been very seriously injured by its ravages; to what extent is vividly shown in the following account by Prof. J. E. Willet, of Macon, Ga. [The following are the essential

Fig. 33.—MELON-WORM (*Phacellura hyalinatalis*).
Larva, Chrysalis, and Moth, closed and open.

points of Prof. Willett's letter. He thus describes the appearance in three patches, in which melons had been planted for market.—ED.]

"All presented the same scene of total destruction. Most of the vines had been more or less denuded of leaves, and the remains of the leaves contained brown chrysalids or *pupæ* "webbed up" in them. The melons of various sizes were occupied in great measure by the worms. The younger worms were generally confined to the surface, but the older had penetrated to different depths. Some had excavated shallow cavities half an inch to an inch in diameter, and one-eighth of an inch in depth; and each cavity was occupied by one or more worms. Others had penetrated perpendicularly into the melons, frequently beyond sight. None had reached the hollow of the melon, so far as I saw. The worms averaged probably half a dozen to each melon. The melon crops of these three market-gardens were a total loss. Another gardener told me that he had abandoned the culture of melons entirely, because of the ravages of the Melon-worm. Where cultivated in considerable numbers, the August and September crop of melons is very uncertain. The destruction is frequently quite complete, also, in private gardens.

"The Melon-worms are of a light yellowish-green color, nearly translucent, have a few scattered hairs, and, when mature, are about an inch and a quarter in length. They "web up" in the leaves of the melon, or of any plant growing near which has flexible leaves, forming a slender brown chrysalis three-quarters of an inch in length. Hundreds of these pupæ were found rolled up in the leaves of the tomato and of the sweet-potato.

"In passing through one of the patches referred to, numbers of small, beautiful moths rose from the grass and weeds. Their wings when extended measured an inch across, and were of an iridescent pearly whiteness, except a narrow black border. Their legs and bodies presented the same glistening whiteness, and the abdomens terminated in a curious movable tuft of white appendages like feathers, of a pretty buff color, tipped with white and black. These moths proved to be the mature melon-worms, which had emerged from the chrysalids referred to.

"The melon-worms, their chrysalids, and moths, were forwarded to Prof. J. H. Comstock, Entomologist of the United States Agricultural Department, for indentification. He pronounced them to be *Phacellura hyalinatalis*, another species of the same genus as the Western pickle-worm, *Phacellura nitidalis*. The moth of the latter is somewhat smaller, and the ground color of the wings is a bronze yellow and the black border is broader.

"Much later in the season a few worms were found on cucumbers, and were pronounced by Professor Comstock to be melon-worms. A year previous, in the summer of 1878, I found a chrysalis webbed in a tomato leaf, and this chrysalis gave forth the same moth, as was found in 1879 to issue from the melon-worm chrysalis. This worm had probably fed on the foliage of a pumpkin vine which ran near the tomato plant.

"The melon-worm, *Phacellura hyalinatalis*, is known then to destroy musk-melons, cucumbers, and pumpkins. Its cousin, the pickle-worm, *Phacellura nitidalis*, has been found here, but it remains to be determined whether it plays any part in the destruction of melons or of cucumbers in this locality.

"No efficient remedy for this has been discovered here. Some have tried placing each melon on a piece of plank, under the mistaken notion that the worms emerged from the earth. Paris green and London purple are objectionable, by reason of their poisonous properties. Professor Comstock has suggested to me a trial of the Persian insect-powder, Pyrethrum. Whatever remedy is employed it must be applied to the leaves as well as to the melons. The worms devour both foliage and fruit, and, if the fruit alone be protected, the foliage will be destroyed, the plants will cease to grow, and the melons will not come to maturity."

The number of broods in a season has not been definitely ascertained. The insect winters in the chrysalis state, spun up in the leaves of any neighboring tree or plant. They usually migrate to a greater or less distance from their feeding place before webbing up. At Rock Ledge, Fla., they were found abundantly webbed up on Palmetto and Orange trees in a grove in which the so-called Indian pumpkins had grown.

As regards remedies, Mr. J. S. Newman, of Atlanta, Ga., states that the only one known to him is to plant early, the object being to pick the melons before the most destructive brood of the worms has appeared. It would undoubtedly be found profitable to keep a sharp lookout for the first brood of the worms, which will probably be found feeding upon the leaves and stems before the young melons have begun to form. These should be killed by hand. This could be readily done in patches comparatively small in size, and we think will be found profitable in large gardens.

Two species of parasitic insects have been reared from the specimens sent to the Department; one is *Pimpla conquisitor*, an Ichneumon fly, which has proved very efficacious in the case of the cotton-worm; the other is a Tachina fly. Much is to be expected from the aid of these parasites.

THE ONION.

THE BLACK ONION-FLY.

(*Ortalis flexa*, Wied.)

This insect was first described by Wiedemann in 1830. The fly (fig. 34) is about one-third of an inch in length, black, with three oblique white stripes on each wing. Mr. Henry Shimer, of Mount Carroll, Ill., says, "In the latter part of June, I first observed the larva or maggot among the onions here. The top dead, the bulb rotten, and the maggots in the decayed substance. From them I bred the fly. They passed about two weeks in the pupa state. At that time I first observed the flies in the garden, and now a few are to be found. Their favorite roosting place is a row of asparagus running along the onion-ground, where they are easily captured and destroyed from daylight to sunrise, while it is cool and wet. During the day they are scattered over the ground and on the leaves and stalks of the onions, and not easily captured. Their wings point obliquely backward, outwards and upwards, with an

Fig. 34.—BLACK ONION-FLY (*Ortalis flexa*) Larva and Fly—real size shown by lines.

irregular jerking, fanlike movement; flight not very rapid or prolonged. They are not very numerous, probably not over two or three hundred. All that I observed originated in one part of the bed, where they were doubtless deposited by one parent fly." Two broods appear in a season.

THE IMPORTED ONION-FLY.

(*Anthomyia ceparum*, Bouché.)

The engravings (fig. 35,) show, *a*, larva; *b*, larva magnified; *c*, pupa; *d*, pupa magnified; *e*, fly magnified. It is a terrible pest to the onion grower in the East, though it has not yet made its way out West. On the other hand, the Native American Onion-fly (*Ortalis arcuata*, Walker), which is a closely allied species and has almost exactly the same habits, has only been heard of in one or two circumscribed localities in the West, and even there does comparatively but little damage.

The Imported Onion Fly lays her eggs while the onions are small (in May and June), depositing them on the leaves near the surface of the ground. The maggots soon hatch and make their way down to the base of the young bulb, sometimes as many as two or three in a single onion: here they feed for about a fortnight, when they usually leave the bulb and turn into chestnut-colored pupæ in the earth near by. In about two weeks the

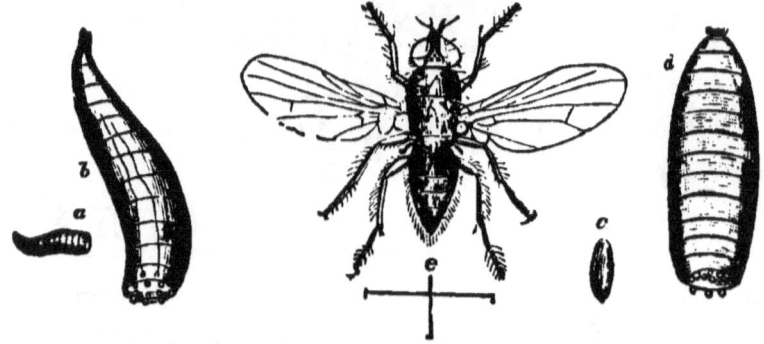

Fig. 35.—IMPORTED ONION-FLY (*Anthomyia ceparum*).
a, Larva; *b*, do. magnified; *c*, Pupa; *d*, do. magnified; *e*, Fly, enlarged, real size shown by the lines below.

second brood of flies issues from these pupæ, to lay eggs and continue the destructive work. The eggs of the later broods are not usually deposited on the leaves, but on the bulb itself, close to the ground.

REMEDIES.—The onions attacked soon turn yellow, and should be removed from the bed at once; this, if thoroughly done, will get rid of the maggots that would develop into the next brood of flies. The sickly onions should be lifted with the aid of an old knife, to be sure and bring up the maggot with the onion; if they are simply pulled, the maggot may escape from the decayed bulb. Place these infested onions in a pail or other vessel from which the maggots can not escape, and burn them. Among special applications, soot has been found useful, and in England kerosene oil mixed with water in the proportion of half a pint to six gallons has been applied to the plants from a watering-pot with a fine rose. Salt applied when the plants are three or four inches high, at the rate of three bushels to the acre, has been used by some of the Connecticut onion growers with benefit.

PARSLEY AND RELATED PLANTS.

In July, in the New England and Middle States, and earlier further South, there will be found upon Parsley especially, and sometimes upon other cultivated umbelliferous plants, as the Carrot, Parsnip, and Celery, and also on Caraway, Fennel, etc., a showy caterpillar, known as the "Parsley-worm." This, when full grown, is an inch and a half long, largest near the head and tapering behind; at this time the caterpillars are of a delicate apple-green color, paler at the sides, and on each segment, or ring of the body, is a band consisting of alternate bright-yellow and black spots. This coloring would be sufficient to identify the caterpillar, but if disturbed it at once arrests attention by appealing to the sense of smell, as it gives off, what has been called a "scent," but is better described as a stench, which pervades the air for some distance. This odor comes from a pair of soft, orange-colored horns, which are united below like a letter Y. These are projected from a slit just back of the head, and are not, as many have supposed, stings, but merely organs for diffusing this odor. They attain their full size late in September or early in October, when they seek some sheltered place on a fence or a building; hang themselves by a loop of silken threads and form greenish, yellowish, or ash-gray chryalids. They pass the winter in the chrysalis state, and the next summer appear as handsome "swallow-tailed" butterflies. The butterfly has a spread of wing of three-and-one-half to four inches. The wings are black, with a row of yellow spots across them, and another row near the margin; the hind wings have each a tail-like appendage, seven blue spots between the two rows of yellow ones, and at the inner angle, an orange-colored spot with a black center. The female lays her eggs singly.

REMEDIES.—The caterpillar is most destructive upon Parsley and the related plants when grown for seed. They devour not only the foliage, but seem to be especially fond of the flower-clusters, and of the young fruit or seeds. As with all other large and scattered caterpillars, hand-picking is the most effective remedy. The butterfly is so handsome that it would not be supposed to be capable of mischief, but seed growers should encourage entomologists to make specimens of all they find.

THE PEA.

THE PEA-WEEVIL.

(*Bruchus pisi*, Linn.)

Our common garden Pea has not many insect enemies, for with the exception of the Striped Flea-beetle (*Haltica striolata*), which gnaws numerous small holes in the leaves, and the Corn-worm, *alias* Boll-worm (*Heliothis armigera*), which eats into the pod, there are very few others besides the Pea-weevil under consideration. This species alone is so numerous, however, as to be a serious drawback to pea culture in some parts of the country.

The Pea-weevil, which is here illustrated (fig. 36),

showing a back view, and (fig. 37), a side view, the small outlines at the sides showing the natural size, is easily distinguished from all other species of the genus with which we are troubled, by its larger size, and by having on the tip of the abdomen, projecting from the wing-covers, two dark oval spots, which cause the remaining white portion to look something like the letter T. It is about 0.18—0.20 inch long, and its general color is rusty-black, with more or less white on the wing-covers, and a distinct white on the hinder part of the thorax, near the scutel. There is a notch on each lateral edge of the thorax, and a spine on the underside of the hind thighs near the apex. The four basal joints of the antennæ, and the front and middle-shanks, and feet, are more or less tawny. It is supposed to be an indigenous North American insect, and was first noticed many years ago around Philadelphia, from whence it has spread over most of the States where the pea is cultivated. This supposition is probably the correct one, though we have no means at present of proving it to be so, and certain it is that, as the cultivated pea was introduced into this country, our Pea-weevil must have originally fed on some other indigenous plant of the Pulse family.

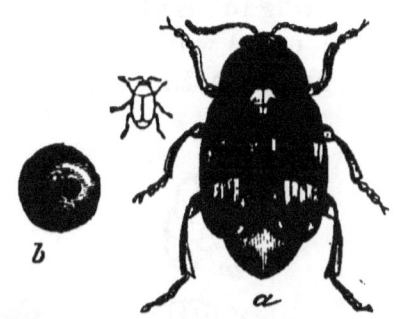

Fig. 36.—PEA-WEEVIL (*Bruchus pisi*). *a*, Back View; *b*, Pea with spot.

THE FEMALE DEPOSITS HER EGGS ON THE OUTSIDE OF THE POD.—It is a very general remark that peas are "stung by the bug," and the impression prevails almost universally, not only among gardeners, but with many entomologists, that the female weevil punctures and deposits her eggs in the pea, in which the larva is to be nourished. It is a little singular that so many writers should have fallen into this error, for it is not only the accepted view with some writers for the agricultural press, but has been adopted by many eminent entomologists. All this comes of course from one man's palming off the opinions of another as his own, whether good or bad, without due credit.

The true natural history of the Pea-weevil may be thus briefly told. The beetles begin to appear as soon as our peas are in bloom, and when the young pods form, the female beetles gather upon them, and deposit their eggs on any part whatever of the surface, without attempting to insert the eggs within the pod.

The eggs, which are of a deep yellow, 0.035-inch long, are three times as long as wide, pointed in front,

Fig. 37.—PEA-WEEVIL.
b, Side View; c, Larva; d, Pupa.

blunt behind, but larger anteriorly than posteriorly. They are fastened to the pod by some viscid fluid, which dries white, and glistens like silk. As the operation of depositing is only occasionally noticed during cloudy weather, we may safely assume that it takes place for the most part by night. If pea vines are carefully examined any time during the month of June, the pods will often be found to have from one to fifteen or twenty such eggs upon them, and the black head of the future larva may frequently be noticed through the delicate shell.

As already stated, the eggs are deposited on all parts of

the pod, and the mother beetle displays no particular sagacity in the number which she consigns to each, for I have often counted twice as many eggs as there were young peas, and the larvæ from some of these eggs, would of course have to perish, as only one can be fully developed in each pea. The newly hatched larva is of a deep yellow color, with a black head, and it makes a direct cut through the pod into the nearest pea, the hole soon filling up in the pod, and leaving but a mere speck, not so large as a pin hole, in the pea. The larva feeds and grows apace, and generally avoids the germ of the future sprout, perhaps because it is distasteful, so that most of the buggy peas will germinate as readily as those that have been untouched. When full grown, this larva presents the appearance of figure 37, *c*, and with wonderful precognition of its future wants, eats a circular hole on one side of the pea, and leaves only the thin hull as a covering. It then retires, and lines its cell with a thin and smooth layer of paste, pushing aside and entirely excluding all excrement, and in this cell it assumes the pupa state, (fig. 37, *d*,) and eventually becomes a beetle, which, when ready to issue, has only to eat its way through the thin piece of the hull, which the larva had left covering the hole. It has been proved that the beetle would die if it had not during its larval life prepared this passage way, for Earnest Menault asserts that the beetle dies when the hole is pasted over with a piece of paper, even thinner than the hull itself.

REMEDIES AND PREVENTIVES.—Sometimes, and especially when the summer has been hot and prolonged, many of the beetles will issue from the peas in the fall of the same year that they were born, but as a more general rule they remain in the peas during winter, and do not issue till new vines are growing. Thus many yet remain in the seed peas until they are planted, and especially is this apt to be the case with such as are planted early. We see, therefore, how easily this insect may be introduced into districts previously free from it, by the careless planting of buggy peas, for it has been demonstrated that the beetle issues as readily from peas planted in the earth, as it does from those stored away in the bin. All peas intended for seed should be examined, and it can very soon be determined whether or not they are infested. The thin covering over the hole of the peas that contain weevils, and which may be called the eye-spot, is generally somewhat discolored, and by this eye-spot, those peas which ought not to be planted, can soon be distinguished. Where this covering is off, and the pea presents the appearance of fig. 36, *b*, there is little danger, for in that case the weevil has either left, or, if still within the pea, is usually dead. It would of course be tedious to carefully examine a large lot of peas, one by one, in order to separate those that are buggy, and the most expeditious way of separating the sound from the unsound, is to throw them into water, when the sound ones will mostly sink, and the unsound swim.

There are, however, other and more certain means of preventing the injuries of this insect, and whenever agriculture shall have progressed to that point, where by proper and thorough organization, all the farmers of a county or of a district can, by vote, mutually agree to carry out a measure with determination, and unison, then this insect can soon be exterminated; for it is easy to perceive that such a result would be accomplished by combinedly ceasing to cultivate any peas at all, for one single year! Until some such united action can be brought about, we shall never become entirely exempt from this insect's depredations, for no matter how sound the peas may be that I plant, my vines are sure to be more or less visited by the beetles, as long as I have slovenly neighbors. Yet, comparatively, my peas will always be enough better to pay well for the trouble, even under these circumstances.

RADISH.

The same insects that attack young cabbage plants and the turnips infest the Radish. In some localities it is almost impossible to grow radishes of a size fit for the table before they are so much injured by a small maggot as to be useless. These maggots appear to be the larvæ of a fly (*Anthomyia*), closely related to those so destructive to the onion. (See Onion Flies). The False Chinch Bug (*Nysius devastator*), troublesome in some of the Western States, attacks the leaves of the Radish as well as those of other plants of the same family.

SQUASH AND PUMPKIN.

The Squash and Pumpkin belong to the same family of plants (the Gourd Family, *Cucurbitaceæ*) with the Cucumber and Melon, and most of the insects that infest those may often be found upon them, especially while the plants are young, at which time they need the protection from the Striped-beetle, etc., mentioned under Cucumber and Melon.

THE SQUASH-BUG.

(*Anasa* [formerly *Coreus*] *tristis*, Degeer.)

For this insect the name, Squash Bug, is scientifically correct, as it belongs to the true bugs, with the Chinch, Bed, and other unpleasant bugs. (See *Hemiptera*, in

Introduction.) About the last of June (in the Northern States), these insects come out from their hiding places, pair, and lay their eggs. The parent insect (fig. 38,) is a little over half an inch ($^6/_{10}$) in length, rusty-black above, and ochre-yellowish beneath. The ground color of the upper parts is ochre-yellow, but concealed by multitudes of minute black dots. A marked character of this insect is the odor it gives off when handled or disturbed; this odor has been compared to "that of an over-ripe pear," but we have never seen a pear sufficiently "over-ripe" to approach in its repulsiveness the sickening odor given off by the Squash-bug. It is one of those odors of which a very little satisfies. The insects are quiet during the day, but at night lay their eggs in little patches; they are of a brownish-yellow color, and glued to the leaves. They soon hatch, and the larvæ, or young bugs, are of a pale-ash color, and of a more rounded shape than the perfect insects. As they grow older they moult their skins several times, forming no dormant pupæ, but finally assume the shape of the perfect bug. The young at first remain in small swarms or clusters, near the place they were hatched, but finally scatter to other leaves; in all stages they penetrate the leaves with their beaks, live upon their juices, and cause them to become brown, wither, and finally to die. As soon as a leaf is exhausted, they pass on to fresher ones, and where numerous, the insects are very destructive.

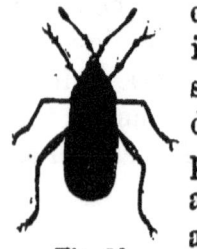

Fig. 38.
SQUASH-BUG.
(*Anasa tristis*.)

REMEDIES.—None of our injurious insects is more readily kept under control than this by hand-picking. If one familiar with the appearance of the bug, will examine the young vines and the ground beneath them, those that come from their winter quarters may be found and destroyed before they have laid their eggs. Next the eggs, which are conspicuous upon the underside of the leaves, on account of their color and large size, may be crushed between the thumb and finger. If any of the patches of eggs have been allowed to hatch, the habit of the young bugs to stay around in clusters, allows these to be readily destroyed. If any have escaped destruction in these early stages, an examination of the vines, after the insects have scattered, will allow of their being gathered and destroyed before they can go into winter quarters.

THE 12-SPOTTED SQUASH-BEETLE.

(Diabrotica 12-punctata, Fabr.)

This beetle is own brother to the Striped-Beetle (*D. vittata*), already described under Cucumbers. While the colors are much the same, this (see fig. 30, p. 45) is much broader, and instead of having its black marks in lines, they are in dots, 12 in all, upon the wing covers. It is, fortunately, not very common, but where it occurs, it is most destructive. It seems to take special delight in eating through the strong ribs of the leaves, near where they are attached to the leaf-stalk, thus allowing the rest of the leaf to fall down and wither. Hand-picking has hitherto been relied upon, but if in great numbers, Paris Green may be used.

THE SQUASH-BORER.

(*Ægeria* [*Trochilium*] *cucurbitæ*.)

At midsummer, or soon after, in the Eastern States, especially, large and vigorous Squash-vines are seen to suddenly wilt and die without apparent cause. Upon a careful examination of the vine near the root, by splitting it lengthwise, there will generally be found a caterpillar, and if the exterior of the vine be carefully examined, probably the wound caused by the entrance of the young borer may be discovered. This Squash-vine Borer is the larval state of an insect of the same genus as the borer of the Peach Tree. The perfect insect (fig. 39,) has an orange-colored body; its fore-wings are black, and the hind ones transparent, and the hind pair of legs are fringed with long orange and black hairs. The female deposits her eggs upon the vine near the root, at any time from June to August. The young larva at once penetrates to the interior of the stem, and eats and grows, until the connection between the upper part and the root being destroyed, the vine dies. The full-grown larva enters the earth, forms a rude cocoon by gluing particles of earth together, and remains in the pupa state until time to begin its work of mischief the next season.

Fig. 39.—MOTH OF SQUASH-VINE BORER.

REMEDIES.—The difficulty with the Squash-vine Borer consists in the fact that its presence is not made known by the wilting of the vines until the mischief has been done. Among other methods that have been suggested, is the placing sheets of the sticky fly-paper ("Catch'em Alive, Oh!") about the vines, to capture the parent insect. If these are seen flitting around the vines, they should be caught by means of a net. If the moths have been seen around the vines, these should be closely examined for eggs and for the wounds made by the young larvæ in entering the stem; if found, while still young, they may be carefully cut out, without material injury to the vine. If the vine dies from the presence of borers, search should be made for the larva, that it may be destroyed, and prevent an increase. Among preventives it has been suggested that, as the insect deposits her egg upon the stem near the root of the vine, the covering slightly with earth of several lower joints of the plants will be effective. A similar treatment has been found useful with the related Peach-tree Borer, and is worth bearing in mind if the Squash Borer is apprehended.

THE TOMATO.

The Tomato, belonging to the same botanical family as the Potato, is attacked by several of the insects that feed upon that plant, and it is not necessary to give a separate description of them. When the plants are first set out in the spring, they are sometimes cut off by the greasy Cut-worm, the larva of *Agrotis telifera*, Harris. This cut-worm is a general feeder, and destroys whatever plants it may come across. The holes of the worms should be searched for, and the tenants destroyed. Wrapping a piece of paper around the lower part of the stem of each plant, allowing the lower edge to be below the surface of the earth, while the other edge extends an inch or two above it, will prevent their attacks.

The Stalk Borer of the Potato, and the Colorado Beetle, occasionally attack the Tomato. These are described under Potato. The most injurious insect to the Tomato, is the large green Caterpillar, of *Sphinx quinque maculata*, which is called both "Tomato," and "Potato Worm," (See POTATO). This voracious feeder will soon strip a plant of its foliage, and even eat the young fruit. Where tomatoes are trained to a trellis, as they always should be in garden culture, the abundant droppings upon the ground will indicate its presence, and it should be sought for at once. Being so nearly of the same color as the stems of the plant, it might escape

notice, did not its droppings betray it. Hand picking—
and it is perfectly harmless—is the remedy.

The Caterpillar of the Corn, or Boll-worm (*Heliothis armigera*), besides doing vast injury to Indian Corn, and to Cotton, feeds on many other plants. In some of the

Fig. 40.—BOLL-WORM (*Heliothis armigera*) FEEDING UPON TOMATO.

Western States, it has proved a great pest to the Tomato grower, eating into the green fruit, and causing it to rot. Figure 40 shows this Caterpillar attacking the Tomato. This insect is described under Indian Corn.

Insects Injurious to Root Crops and Indian Corn.

BEET AND MANGEL WURZEL.

While there are several insects injurious to the Beet, especially the Sugar Beet in Europe, this crop has been thus far singularly free from insect enemies in this country. Even the all-devouring Western Grass-hopper often leaves the Beets untouched; while the White Grub (*Lachnosterna*), makes no such exception, and is sometimes very troublesome. (SEE WHITE GRUB). In England, the larva of the Beet Carrion Beetle (*Silpha opaca*, Linn.), has occasionally destroyed crops by feeding on the leaves, but little seems to be known about it. A fly (*Anthomyia betæ*, Curtis), is often destructive to the Beet and Mangel in Europe, and a few cases have been reported of its appearance in this country in 1881. This insect is a near relative of the Onion Fly (See ONION), its larvæ burrowing in numbers in the pulpy matter of the leaf. When a leaf wilts, the larvæ, about a third of an inch long, may be seen, if present, by holding the leaf up to the light. The only remedy thus far suggested is, to remove all leaves that show signs of flagging, and destroy them before the maggots can transform to flies.

INDIAN CORN.

While there are a few insects that especially attack this important crop, it also receives attention from the generally destructive insects. In those localities where the

Army Worm is abundant, or in those that are in the range of the destructive Western Grass-hopper, corn, of course, suffers in common with other plants. It also is attacked by the White Grub, and by several of the Cut Worms.

THE CORN-WORM, *alias* BOLL-WORM.

(*Heliothis armigera*, Hubner.)

The "Boll-worm" has become a by-word in all the Southern cotton-growing States, and the "Corn-worm" is a like familiar term in those States, as well as in many

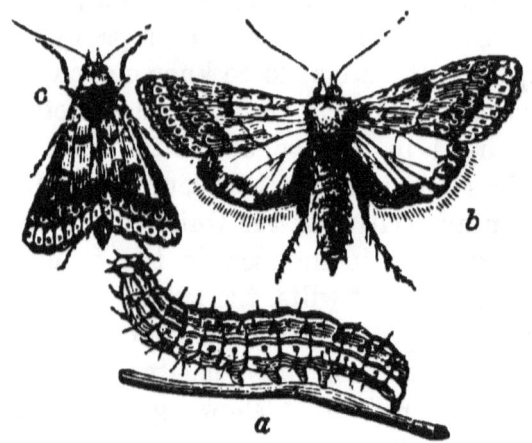

Fig. 41.—CORN OR BOLL WORM.
a, Larva; *b*, Moth, Wings open; *c*, Moth, Wings closed.

other parts of the Union; but few persons suspect that these two worms—the one feeding on the corn, the other on the cotton-boll—are identically the same insect, producing exactly the same species of moth. But such is the fact, as I myself experimentally proved in 1864. It attacks corn in the ear, at first feeding on the "silk," but afterwards devouring the kernels at the terminal end; being securely sheltered the while within the husk. I have seen whole fields of corn nearly ruined in this way, in the State of Kentucky, but nowhere have I known it to be so destructive as in Southern Illinois. Here, as in our own State, there are two broods of the worms during the year, and very early and very late corn fare the worst; moderately late and moderately early varieties usually escaping. I was formerly of the opinion that this worm could not live on hard corn, and it certainly does disappear before the corn fully ripens, but last fall Mr. James Harkness, of St. Louis, brought me, as late as the latter part of October, from a corn field on the Illinois bottom, a number of large and well ripened ears, each containing from one to five worms of different sizes, subsisting and flourishing on the hard kernels. This is, however, an exceptional occurrence, brought about, no doubt, by the long protracted warm weather which we had, and the worms were in all probability a third brood.

This glutton is not even satisfied with ravaging these two great staples of the country—cotton and corn—but, as I discovered, in 1867, it attacks the tomato in Southern Illinois, eating into the green fruit, and thereby causing such fruit to rot. (See TOMATO, p. 66). Mr. Glover also found it feeding in a young pumpkin, and it has been ascertained by Mrs. Mary Treat, of Vineland, New Jersey, not only to feed upon the undeveloped tassels of corn and upon green peas, but to bore into the stems of the garden flower known as Gladiolus, and in confinement to eat ripe tomatoes, last summer it was also found by Miss M. E. Murtfeldt on common string beans, around Kirkwood.

But for the present we will consider this insect only in the role of Corn-worm, because as such it interests the practical man most deeply.

This insect is very variable in the larva state, the young worms varying in color from pale-green to dark-brown. When full grown there is more uniformity in this respect, though the difference is often sufficiently great to cause them to look like distinct insects. Yet the same pattern is observable, no matter what may be the general color; the body being marked as in figure 41, with longitudinal light and dark lines, and covered with black spots which give rise to soft hairs. Those worms that Mrs. Treat found on green peas and upon corn tassels, had these lines and dots so obscurely represented that they seemed to be of a uniform green or brown color, and the specimens which I saw last summer on string beans were also of a dark glass-green color, with the spots inconspicuous, but with the stripe below the breathing pores quite conspicuous and yellow. The head, however, remains quite constant and characteristic. Figure 40 may be taken as a specimen of the light variety, and figure 41, *a*, as illustrating the dark variety. When full grown, the worm descends into the ground, and there forms an oval cocoon of earth interwoven with silk, wherein it changes to a bright chestnut-brown chrysalis, provided with four thorns at the extremity of the body, the two middle ones being stouter than the others. After remaining in the chrysalis state from three or four weeks, the moth makes its escape. In this last and perfect stage, the insect is also quite variable in depth of shading, but the more common color of the front wings is pale clay-yellow, with a faint greenish tint, and they are marked and variegated with pale-olive and rufous, as in figure 41 (*b* showing the wings expanded, and *c* representing them closed), a dark spot near the middle of each wing being very conspicuous. The hind wings are paler than the front wings, and invariably have along the outer margin a dark brown band, interrupted about the middle by a large pale spot.

In 1860—the year of the great drouth in Kansas—the corn crop in that State was almost entirely ruined by the Corn-worm. According to the "Prairie Farmer," of January 31, 1861, one county there which raised 436,000 bushels of corn in 1859, only produced 5,000 bushels of poor wormy stuff in 1860, and this, we are told, was a fair sample of most of the counties in Kansas. The damage done was not by any means confined to the grain actually eaten by the worm; but "the ends of the ears of corn, when partially devoured and left by this worm, affords a secure retreat for hundreds of small insects, which, under cover of the husk, finish the work of destruction commenced by the worm, eating holes in the grain or loosening them from the cob. A species of greenish-brown mould or fungus grew likewise in such situations, it appearing that the dampness from the exuded sap favored such a growth. Thus decay and destruction rapidly progressed, hidden by the husk from the eye of the unsuspecting farmer." It appears also that many horses in Kansas subsequently died from disease occasioned by eating this half-rotten wormy corn.

REMEDY.—It is the general experience that this worm does more injury to very early and very late corn than to that which ripens intermediately, for though the broods connect by late individuals of the first and early individuals of the second, there is nevertheless a period about the time the bulk of our corn is ripening, when the worms are quite scarce. I have never yet observed their work on the green tassel, as it has been observed in New Jersey, and do not believe that they do so work with us. Consequently it would avail nothing as a preventive measure, to break off and destroy the tassel, and the only remedy when they infest corn is to kill them by hand. By going over a field when the ears are in silk, the presence of the worms can be detected by the silk being prematurely dry, or by its being partially eaten.

In the cotton fields large numbers of the moths have been caught by means of lamps or lanterns, so arranged that the insect, when attracted by the light, will fall into the water or other liquid. Wherever the moth is abundant among the corn, it may be worth while to try this.

THE SEED-CORN MAGGOT.

(*Anthomyia zeas*, Riley).

This maggot is shown, enlarged, at figure 42, *a*, the line directly underneath giving the natural size. It greatly resembles the Onion maggots, which are known to attack the onion in this country, and its work on corn is similar to that of this last named maggot on the onion; for it excoriates and gnaws into the seed-corn as shown in figure 43, and finally causes such seed to rot.

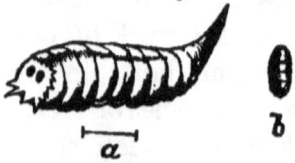

Fig. 42.—SEED-CORN MAGGOT.
a, Larva; *b*, Pupa.

After having become full fed, these maggots usually leave the kernels for the surrounding earth, where they contract into smooth, hard, light-brown pupæ, of the size and form of fig. 42, *b*, and in about a week afterwards the perfect fly pushes open a little cap at the anterior end, and issues forth to the light of day. In this state it is a two-winged fly belonging to the Order *Diptera*, and quite inconspicuous in its markings and appearance.

It is difficult to suggest a remedy for this pest, as its presence is not observed before the mischief is done. Hot water has been found effectual in killing the Onion maggot, without injuring the onions, and would doubtless prove as effectual for this Corn maggot, where a few hills of some choice variety are attacked, which it is very desirable to save.

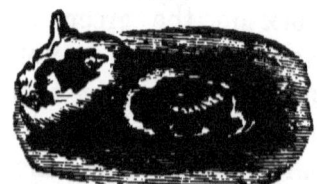

Fig. 43.—MAGGOTS AT WORK.

But its application in a large field, even if one knew where to apply it, would be impracticable, and I can only suggest soaking the seed before planting, in gas-tar or copperas, and hope that the experiment will be tried by those of our Eastern friends who have suffered from this maggot.

Some species are such general feeders that, in grouping insects according to the plants to which they are injurious, it becomes difficult to place them. The corn-grower, the vegetable gardener, the grower of small fruits, the owner of lawns and pastures, and even the nurseryman, may each at times properly look upon the White Grub as his worst insect enemy. The Cut-worms and Wire-worms, though less generally destructive, are injurious to very unlike plants, while the False Wire-worms are in this country, especially, known for the injury they have done to lilies, carnations, and to the potato crop. We place these general feeding insects here, for convenience, and shall refer to them in treating of the plants to which they are especially injurious.

THE WHITE GRUB.

(*Lachnosterna fusca*, Frohl).

Perhaps no destructive insect is better known than this in its larval as well as in its perfect state, by those who live in the country, yet comparatively few are aware that the frequent White Grub and the familiar May-bug, or June-bug, or Dor-bug, are different forms of the same insect. In the months of May and June, attracted by the light, these beetles often make their way into the house, and by the noise they make in buzzing about and knocking themselves against the walls and ceiling, often alarm nervous persons. The few that thus enter the dwelling are merely indications that vast swarms are upon the trees at no great distance without. Unlike some beetles, this is a voracious feeder in its perfect state, and is destructive to trees, sometimes completely denuding them of their foliage, without the cause being discovered, as the beetle is active only at night. Fruit and ornamental trees, as well as forest trees, appear to be attacked indiscriminately. They remain in the beetle state but a short time, and the damage they do is small as compared with that which they inflict in their prolonged grub state. The beetle is about an inch long, of the shape shown in figure 44; its legs are long and slender, with sharp claws, by which it can hold readily to the foliage, etc.; it is of a dark-chestnut color, and covered with minute dots; each wing-cover has two or three slighly elevated longitudinal lines, and the breast is covered with a yellowish down. If the small feelers be examined, the knob at the end will be found to consist of three leaf-like plates.

Fig. 44. JUNE-BUG.

Soon after pairing the female enters the earth to the depth of a few inches, she there deposits forty or fifty eggs, and soon dies. The eggs hatch in about a month, and as the grubs are at first quite small, but little is known of their history during their first year, but they no doubt subsist upon any small roots they may come across. In the second year they are large enough to make their presence felt; they then work near the surface, and it seems to make little difference what kind of root they meet with, it is cut off a short distance below the surface of the ground, and the plant wilts and dies. This happens to Indian corn, to grass, to tender lettuce in the garden, and the woody roots of young fruit trees in the nursery, as well as to the more tender ones of the Strawberry; besides, it often revels in the tubers of the potato, making the crop fit only for the pigs; it also does mischief in the flower garden—indeed, no live root seems to come amiss to this general feeder. The grub is full-grown in the

Fig. 45.—WHITE GRUB.

spring of the third year (some say the fourth), and is then sometimes as large as one's little finger, of the shape shown in figure 45. It is soft, dirty-white, and has a mahogany-colored head, and is usually found with its body curved in a semicircle, though it can straighten itself out and crawl slowly. In the third year they form a somewhat egg-shaped chamber, by sticking the particles of earth together by means of an adhesive fluid, within which they assume the pupa state. These earthen cocoons are shown in figure 46, entire, and in figure 47 cut open, showing the pupa within. In May, or in many localities not until June, the change into the perfect beetle is completed. Such is, in brief, the natural

Fig. 46.—COCOON.

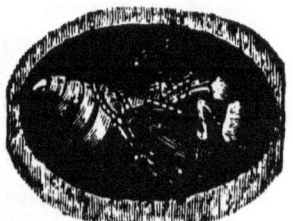

Fig 47.—PUPA.

history of the White Grub, which in most respects resembles that of a related insect, equally destructive to the vegetation of Europe, *Melolontha vulgaris*, known in England as Cockchafer, and in France as Hanneton, which in the last-named country causes such losses that various prizes for efficient means for its destruction have been offered, but not awarded.

It should be stated that a larva of similar size and appearance to the White Grub is often found in manure heaps, and farmers, supposing them to be identical, fear to cause trouble by using the manure. This grub, known as the Muck-worm, is the larva of a different beetle (*Ligyrus*), and as it feeds only upon decayed vegetable matter, can do no damage to the crops; at most it can only consume a little manure. It has a lead-colored appearance, for its whole length, due to the contents of its intestines, which show through the skin; the White Grub shows this dark color only near the tail end.

REMEDIES.—THE BEETLE—As many insects are not injurious in their perfect form, the June-bug has not generally been regarded as harmful. As it is a destructive feeder in its beetle state, it should be destroyed not only for the mischief it may do as a beetle, but for the prevention of its progeny. Those that enter the house should be caught and killed. In each locality the insect is usually more numerous than at other times, once in three years. When the trees in which they harbor are dis-

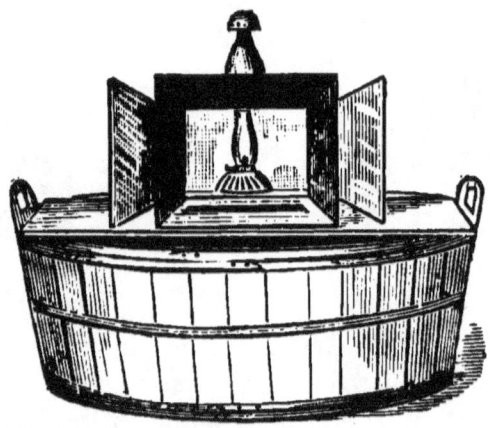

Fig. 48.—LANTERN TRAP FOR THE JUNE-BUG.

covered, large numbers may be destroyed by placing sheets beneath the trees, and in early morning, when they are torpid, the beetles may be shaken from the trees, gathered, and destroyed. As it is only in the perfect state that any effective war can be waged against the White Grub, every method should be employed to kill as many beetles as possible. That which promises to be most effective is, to take advantage of the fact that they are attracted by light, and to set traps for them. A simple form is to place a lamp in a tub, made by sawing a barrel in halves, in the bottom of which is a few inches of water.

Several lanterns have been especially devised which, by throwing a strong light, will attract the insects. That shown in figure 48 is a French device, modified by a distinguished American horticulturist. It consists of a square glass lantern, at each side of which is a flaring tin reflector. At the lower side of each reflector, near the glass, is a longitudinal opening. The lantern is set upon a cask or tub in which there is some water. The insects, attracted by the light, fly towards it, and striking the glass fall down through the opening into the water. This has been found very effective. It would no doubt be well to place upon the water a little kerosene, just a thin film, enough to cover each insect as it falls in.

THE GRUB.—In Europe, with the related grub, the habits of which are like those of ours, the employment of children to follow the plow to pick up the larvæ as they are exposed, has been found the only efficient method. It is probable that pigs and ducks might be made useful in a similar manner.

In gardens, when a vegetable, a fruit, especially the strawberry, or a flower, is observed to suddenly wilt and droop, it should at once be lifted, with the surrounding earth, and the grub sought for and destroyed. The grubs sometimes cut the roots of the grass in lawns to such an extent that the turf may be rolled up like a rug. Of course there is no remedy, but to prevent further trouble the grubs should be picked up, and they are sometimes gathered by the bushel, before re-seeding the lawn. Pastures, in which the grub is present in large numbers, should be given over to the swine, which will soon dispose of them.

NATURAL REMEDIES.—Chief among these is the much abused Crow, which is most efficient in discovering these grubs and destroying them in grass lands; by some instinct they discover the hidden enemy, and many of these birds will regularly visit an infested lawn or pasture every morning and prosecute their beneficial work. It is not unlikely that much of the corn pulling for which these and other birds are blamed is done in the search for these grubs. The Skunk is very fond of the beetles and destroys large numbers of them. In a number of localities in Ohio, Iowa, and Missouri, White Grubs have been found with a long horn protruding from each side of the head, as in figure 49. These "horns" do not properly belong to the grub, but are really vegetable para-

Fig. 49.—WHITE GRUB WITH PARASITIC FUNGUS.

sites, being a kind of fungus. The occurrence of fungi upon other grubs in some Oriental countries has long been known, and the occasional abundance of this upon our native pest, encourages the hope that here may be found an important aid to the cultivator. At all events, grubs found with these horn-like appendages should not be destroyed, but left with the hope that the beneficial vegetable may be propagated and become common.

THE CUT-WORMS.

Among the greatest enemies to the Corn crop, especially in its young state, are the Cut-worms, though their attacks are by no means confined to this plant, but they feed upon a great number of cultivated plants, cutting them off near the surface of the ground. It is a comparatively recent discovery that some of these worms, formerly supposed to feed only on the ground, climb fruit trees and injure their buds. These are mentioned under "Insects Injurious to Fruit Trees." The name Cut-worm is sometimes incorrectly applied to the White Grub, the larva of the May or June-bug, and also to the Wire-worms. The proper Cut-worms are the larvæ of several night-flying moths, of the genus *Agrotis*, and of some allied genera, but as their habits are much alike, a description of one will answer in a general way for all. Related species are destructive in England, where they are known as "Surface Caterpillars."

The moths, which usually appear in late summer, have an expanse of wing of about an inch and a half; they are of a sombre gray or brown color; they rest with the wings closed more or less flatly over the body, the upper entirely covering the lower ones, and always have two, more or less distinctly marked spots, the one round and the other kidney-shaped. The moths, attracted by the lights, frequently enter houses at night; they sometimes fly in cloudy days also. They deposit their eggs mostly in late summer, sometimes in spring, upon plants near the surface of the ground; these soon hatch and the young larvæ enter the earth, where they live upon the tender roots of grass and other plants, until winter, when, about two-thirds grown, they descend deeper into the soil, and remain in a torpid state during cold weather. In spring they come to the surface, and with appetites sharpened by their long fast, are ready to attack almost any succulent plant. They feed by night, and hide in holes just under the surface during the day. They have a general greasy appearance, being smooth, naked, and of some shade of gray, green, brown, or black, and variously marked; the head is polished, and there is a shield of the same color upon the top of the first and last segments; when disturbed they coil themselves into a ball. When the worms are full grown, they descend deeper into the ground, form an earthern cocoon, in which they become chrysalids, and in summer or early autumn appear as moths, to continue the round of changes. There

Fig. 50.
CUT-WORM.

are half a dozen or more species of this terrestrial or non-climbing Cut-worms; the one chosen for illustration (fig. 50), the Greasy Cut-worm (*Agrotis telifera*, Harr.), is one of the most common, and will give a general idea of all. It appears to attack nearly all green cultivated plants with equal avidity, and has proved most destructive to corn, tomatoes, and tobacco.

REMEDIES.—Birds give much aid by destroying the larvæ when exposed by the plow. Chickens will destroy large numbers if cooped in the garden. There are several insect enemies; Ichneumon and other flies deposit their eggs within the worm. Some of the Cannibal beetles, and some spiders prey upon them. Finding their hiding places and killing the worm, is the most effective artificial remedy. The worms usually secrete themselves in the ground near the place where they have destroyed a plant, and often drag a leaf to the entrance of the hole, as if to serve as a guide to it. It has been recommended to make a number of smooth holes near the hill of corn, or near other plants that are attacked, by means of a small stick; many will take refuge in these, and may be killed the next morning by the use of the same stick. This method has been carried out more expeditiously by the use of a circular block of wood, with several smooth pegs in its under side near the edge. This is furnished with a handle, and thrusting it down upon the soil, it will make a series of smooth holes, in which the worms will take refuge; the next day they may be killed by the use of the same implement. In localities where loss from these worms is apprehended, it is a wise precaution to use an abundance of seed, so that a good stand may be left after they have done their mischievous work.

WIRE-WORMS.

The term Wire-worm, like that of Cut-worm, is sometimes applied rather indefinitely; the name properly belongs to the long and slender larvæ of several species of *Elater*, popularly known as "Spring" and as "Click-beetles." A very large blackish beetle, nearly two inches long, with two large round black spots on its thorax which are mistaken for eyes, is often found on fences, sides of buildings, etc., in summer. This, when laid upon its back, will by a sudden spring throw itself into the air to the height of several inches, and usually come down right side up. This, the most conspicuous of these beetles, is not injurious, but there are several others, much smaller, but with the same power of springing, the

Fig. 51.—WIRE-WORM.

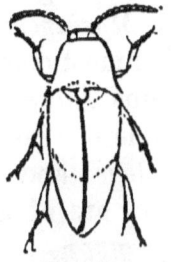

Fig. 52.
SNAP-BEETLE.

larvæ of which are Wire-worms. Entomologists have placed some of these insects in different genera, but for the present purpose, it is sufficient to regard them all as Spring-beetles, and their larvæ as Wire-worms. The larvæ of some live only in decayed wood, and are not injurious, while others live in the soil, and do great damage to several crops. Figure 51 gives the general appearance of the larva, and figure 52 of the beetle, though in some the beetle is much narrower in proportion. So far as is known of their history, these larvæ live for several years in the ground, some say for three, and others for five years. They are all long in proportion to their diameter, their form suggesting the name Wire-worm. Their injury to the Potato crop is perhaps more generally noticed, as it is sometimes completely ruined by them; they also do much damage to Indian Corn, the cereal grains and the grasses. Plowing, both in fall and early spring with frequent harrowing, will expose them to the birds, who are the chief help. In England, previous to planting the potato crop, potatoes, with a stick thrust into them to mark the place, are buried here and there to serve as traps; they are taken up at intervals, and any worms that may have collected on them destroyed.

FALSE WIRE-WORMS.

Several worm-like creatures found in the soil are popularly called wire-worms, which are not the larvæ of the Snap-beetles; indeed are not any kind of a larva. These are now regarded as belonging as to a sub-order of insects, the *Myriapods*, which includes Centipedes, Millipedes, etc. The most common representatives of these belong to the genus *Iulus*. They have worm-like bodies, made up of numerous horny divisions, most of which bear two pairs of legs, and there are two short feelers at the head. They are of a blackish or dark-brown color, and when disturbed, coil themselves into a ring. They undergo no metamorphosis like the proper insects, from which they are also distinguished by their numerous legs. Our species are from an inch to an inch and a half long, but in tropical countries they reach six and seven inches. Many of them feed upon decayed vegetable and animal matter, but some of them feed upon the roots of living plants,

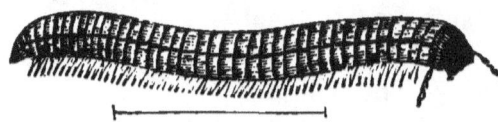

Fig. 53.—FALSE WIRE-WORM (*Iulus*).

The engraving (fig. 53), which illustrates their general appearance, is of our largest species (*Iulus multistriatus*, Walsh.), which has been found in some localities destructive to strawberry plants, carnations, and especially to lily-bulbs. Potatoes have also been much injured by smaller species. Traps in the form of potatoes, as mentioned under Wire-worms, would be of service, or slices of apples, carrots, potatoes, or parsnips, placed upon the beds and covered with pieces of board, will catch many of these millipedes.

THE POTATO.

The late B. D. Walsh, the lamented senior editor of the "American Entomologist," contributed a valuable paper to that journal, of which the following is the substance. After commenting upon the absurdity of the various articles in the papers on "The Potato Bug," he shows that there are a number of insects that are injurious to the potato, and describes the most important, beginning with

THE STALK BORER.

(*Gortynia nitida*, Guenee.)

This larva (fig. 54, 2), commonly burrows in the large stalks of the potato; but is not peculiar to that plant, as it occurs also in the stalks of the tomato, and in those of the dahlia and aster and other garden flowers. We have likewise found it boring through the cob of growing Indian corn, and strangely confining itself to that portion of the ear; and we formerly received a single specimen embedded in the stem of Indian corn, from which we subsequently bred the winged insect. By way of compensation, we suppose, it is particularly partial to the stem of the common Cocklebur (*Xanthium strumarium*); and if it would only confine itself to such noxious weeds as this, it might be considered as a friend instead of an enemy. Fourteen years ago it was more numerous than usual, and we noticed it to be particularly abundant along the Iron Mountain and Pacific roads in Missouri.

The larva of the Stalk Borer moth leaves the stalk in which it has burrowed the latter part of July, and descends a little below the surface of the earth, where in about three days it changes into the pupa or chrysalis state. The winged insect (fig. 54, 1), which belongs to the same

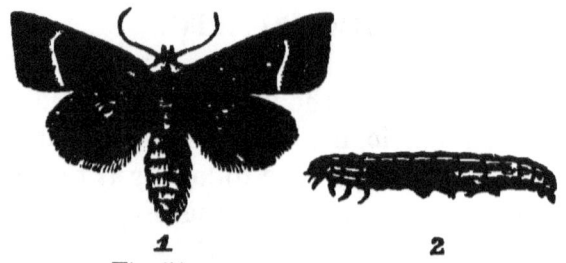

Fig. 54.—POTATO-STALK BORER.
1, Moth; 2, Larva.

extensive group of moths (*Noctua* family, or owlet moths) to which all the cut-worm moths appertain, emerges from under ground from the end of August to the middle of September. Hence it is evident that some few, at all events, of the female moths must live through the winter in obscure holes and corners, and lay their eggs upon the plants which they infest in the following spring. For otherwise, as there are no young potato or tomato plants, or Indian corn, or dahlias, or asters, or even cocklebur for them to lay their eggs upon in the autumn, the whole breed of them would die out in a single year. When a vine is found to wilt suddenly, it should be examined for this insect, which should be destroyed, to prevent further increase.

THE POTATO-STALK WEEVIL.

(*Baridius trinotatus*, Say).

This insect is more peculiarly a southern species, occurring abundantly in the Middle States, and in the more southerly parts of Indiana and Illinois, and also in Missouri; but, according to Dr. Harris, being totally unknown in New England. The female beetle (fig. 55, *c*), deposits a single egg in an oblong slit about one-eighth inch long, which it has previously formed with its beak in the stalk of the potato. The larva subsequently hatches out, and bores into the heart of the stalk, always, according to Miss Morris, of Pennsylvania, who was the first to notice it, proceeding downwards towards the root.

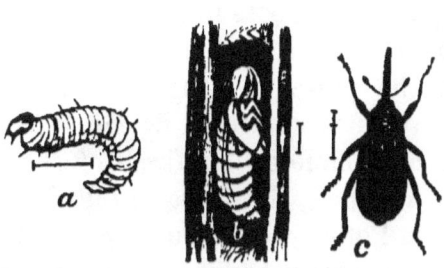

Fig. 55.—POTATO-STALK WEEVIL.
a, Larva; *b*, Pupa; *c*, Weevil.

When full grown, it is a little over one-fourth inch long (fig. 55, *a*), and is a soft whitish, legless grub, with a scaly head. Hence it can always be readily distinguished from the larva of the Stalk Borer, which has invariably sixteen legs, no matter how small it may be. Unlike this last insect, it becomes a pupa (fig. 55, *b*), within the potato stalk which it inhabits; and it comes out in the beetle state about the last of August or the beginning of September. The stalk inhabited by the larva almost always wilts and dies. So soon as the vines first wilt, they should be pulled up and burned. The perfect beetle, like many other snout-beetles, must of course live through the winter to reproduce its species in the following spring.

Miss Morris found that "in many potato fields in the neighborhood of Germantown, Penn., every stem was infested by these insects, causing the premature decay of the vines and giving to them the appearance of having been scalded.

THE POTATO-WORM OR TOMATO-WORM.

(Sphinx quinque-maculata, Haworth).

This well-known insect, the larva of which is usually called the Potato-worm, but it is far commoner on the closely allied tomato, the foliage of which it often clears off very completely in particular spots in a single night. Many persons are afraid to handle this worm, from an absurd idea that it has the power of stinging with the horn on its tail. This worm is shown in fig. 56, about two-thirds grown. We have handled hundreds of them with perfect impunity; in fact, this dreadful looking horn is not peculiar to the Potato-worm, but is met with in almost all the larvæ of the large and beautiful group to which it belongs (*Sphinx* family). It seems to have no special use, but, like the bunch of hair on the breast of the turkey cock, to be a mere ornamental appendage.

When full-fed, which is usually about the last of August, the Potato-worm burrows under ground and shortly afterward transforms into the pupa state (fig. 57. The pupa is very often dug up in the spring from ground where tomatoes or potatoes were grown in the preceding season; and most persons that meet with it suppose that the singular, jug-handled appendage at one end of it, is its tail. In reality, however, it is the tongue-case, and contains the long pliable tongue which the future moth will employ in lapping up the nectar of the flowers, before which, in the dusky gloom of some warm, balmy summer's evening, it hangs for a few moments suspended in the air.

The moth itself (fig. 58), was formerly confounded with the Tobacco-worm moth (*Sphinx Carolina*, Lin-

OF THE FARM AND GARDEN. 87

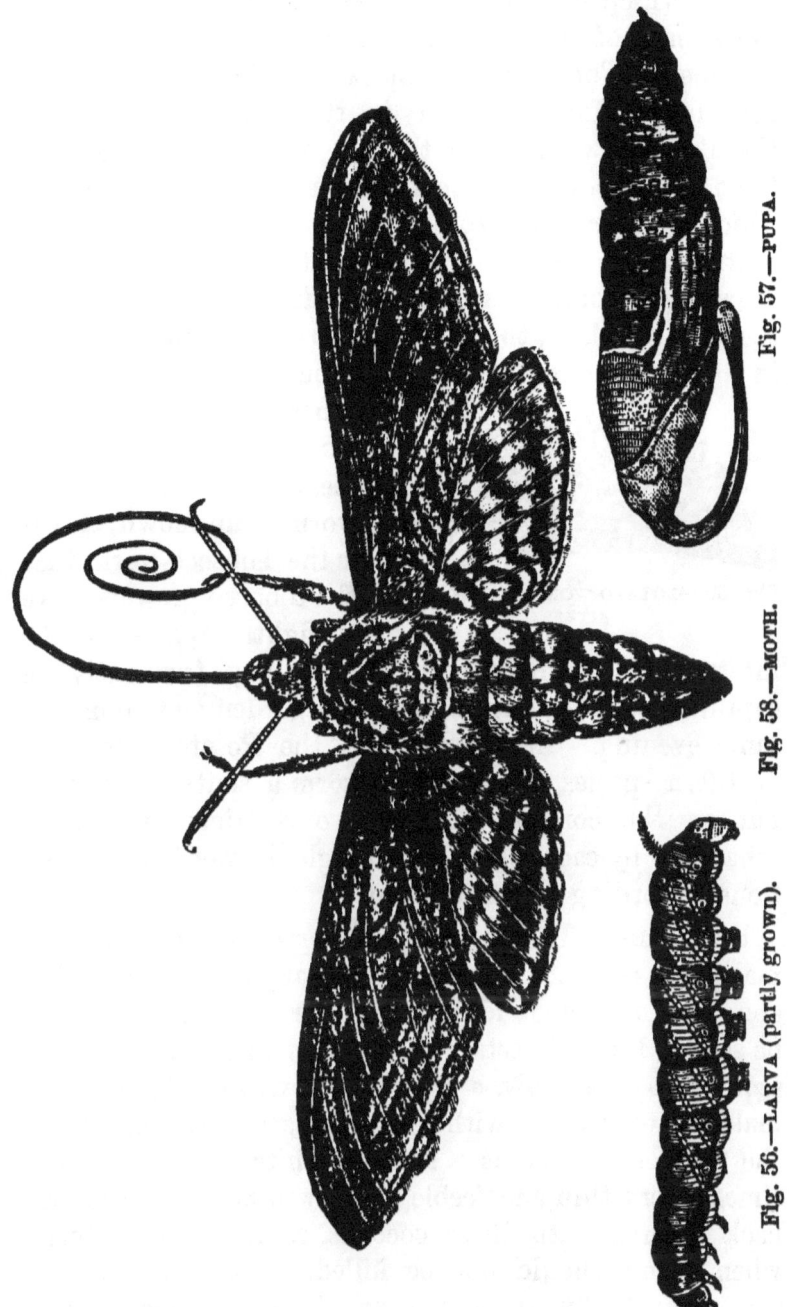

Fig. 57.—PUPA.
Fig. 58.—MOTH.
Fig. 56.—LARVA (partly grown).

næus), which indeed it very closely resembles, having the same series of orange-colored spots on each side of the abdomen. The gray and black markings, however, of the wings differ perceptibly in the two species; and in the Tobacco-worm moth there is always a more or less faint white spot or dot near the centre of the front wing, which is never met with in the other species. In Connecticut and other Northern States where Tobacco is grown, the Potato-worm often feeds upon the leaves of the Tobacco plant, the true Tobacco-worm being unknown in those latitudes. In the more southerly States, on the other hand, and in Mexico and the West Indies, the true Potato-worm is unknown, and it is the Tobacco-worm that the tobacco growers have to fight. While in the intermediate country both species may frequently be captured on the wing in the same garden and upon the same evening. In other words, the Potato-worm is a northern species, the Tobacco-worm a southern species; but on the confines of the two districts exclusively inhabited by each, they intermingle in varying proportions, according to the latitude.

Fig. 59.—POTATO-WORM, WITH PARASITES.

REMEDIES.—The larva is so voracious that it soon makes its presence known by the bare stems, and by the abundant droppings found upon the ground, and should be sought for and destroyed. It has more than one insect enemy, notably a fly, the larva of which, after making its growth within the Potato-worm, comes to the surface and spins a smooth white cocoon. Sometimes a very thin and feeble worm will be found with its back covered with these cocoons, as in fig. 59. Such, when found, should not be killed, as it is desirable to propagate the fly, and the worm will never perfect itself.

Tobacco growers sometimes place some poisonous syrup in the long tubular flowers of the Jamestown Weed (*Datura Stramonium*), and thus kill the moths.

THE STRIPED BLISTER-BEETLE.

(*Lytta vittata*, Fabr).

The three insects just described infest the potato plant in the larva state only, the first two of them burrowing internally in the stalk or stem, the third feeding upon its leaves externally. Of these three the first and third are moths or scaly-winged insects (Order *Lepidoptera*). The second of the three, as well as the next four foes of the potato, which we shall notice, are all of them beetles or shelly-winged insects (Order *Coleoptera*). As these four species all agree with one another in living under ground and feeding upon various roots, during the larva state, and in emerging to attack the foliage of the potato, only when in the course of the summer they have passed into the perfect or beetle state, it will be quite unnecessary to repeat this statement under the head of each of the four. In fact, the four are so closely allied, that they all belong to the same family of beetles, the Blister-beetles (*Lytta* family)—to which the common imported Spanish-fly or Blister-beetle of the druggists appertains—and all of them will raise just as good a blister as that does, and are equally poisonous when taken internally in large doses. The Striped Blister-beetle (fig. 60,) is almost exclusively a southern species, occurring in particular years very abundantly on the potato vine in Central and Southern Illinois, and also in Missouri, but in North Illinois being usually rare. A few years ago it was reported by Mr. Graham Lee, of Mercer County, of N. Ill., and also

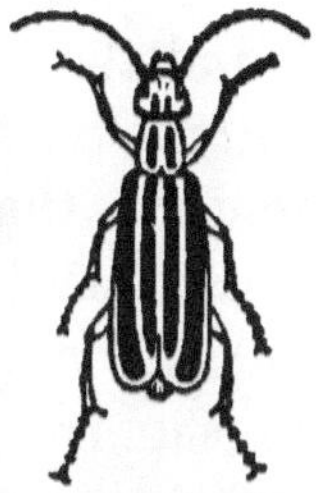

Fig. 60.—STRIPED BLISTER-BEETLE.

by Capt. Beebe, of Galena, N. Ill., as occurring in very large numbers upon their potatoes, and, according to Dr. Harris, it is occasionally found even in New England. In some specimens, the broad outer black stripe on the wing-cases is divided lengthways by a slender yellow line, so that instead of two there are three black stripes on each wing-case; and in the same field we have noticed, on two separate occasions, that all the intermediate grades between the two varieties may be met with; thus proving that the four-striped individuals do not form a distinct species, as was formerly supposed, but are mere varieties of the same species to which the six-striped individuals appertain. Some years since we found the insect very abundant on the potato in Champaign Co., Ill., and Mr. Merton Dunlap, of Champaign, told us that he had succeeded in driving them with brush off his potato-patch on to some old hay which he had prepared to receive them, and then, setting fire to the hay, consumed them bodily. Many such cases may be found recorded in different agricultural journals.

Mr. M. S. Hill, of East Liverpool, Ohio, states in the "Practical Entomologist" (vol. I, p. 197), that this species had once swarmed on the potato vines in his neighborhood, and that "the most successful method of destroying them was by placing between the furrows or rows, dry hay or straw, and setting it on fire." "The bugs," he adds, "were thus nearly all destroyed, and the straw burning very quickly did not injure the vines."

THE ASH-GRAY BLISTER-BEETLE.

(*Lytta cinerea*, Fabr.)

This species (fig. 61, *a*, male) is the one commonly found in the more northerly parts of the Northern States, where it usually takes the place of the Striped Blister-beetle. It is of a uniform ash-gray color; but this color is given it by the presence upon its body of minute ash-gray scales or short hairs, and whenever these are rubbed off, which happens almost as readily as on the wings of a butterfly, the original black color appears. It attacks not only potato vines, but also Honey-locusts, and especially the English and Windsor bean. In one particular year, we have known them, in conjunction with about equal numbers of the common Rose-bug (*Macrodactylus subspinosus*, Linn.), to swarm upon every apple tree in a small orchard in Northern Illinois, not only eating the foliage, but gnawing into the young apples. They were formerly quite common in parts of Illinois, Missouri, Wisconsin,

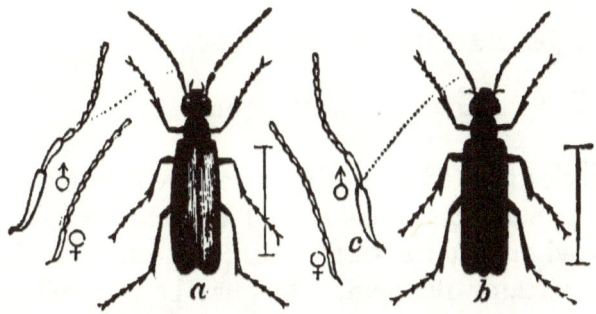

Fig. 61.—ASH-GRAY (*a*), AND BLACK-RAT (*b*) BLISTER-BEETLES.

and Iowa; and the people there got so habituated to the presence of the Colorado bug, that in many cases they thought it was a fresh invader from the region of the Rocky Mountains: whereas it has existed everywhere in the more northerly parts of the United States from time immemorial.

THE BLACK-RAT AND BLACK BLISTER-BEETLES.

(*Lytta murina*, Leconte, and *L. atrata*, Fabr.)

The first of these, the Black-rat Blister-beetle (fig. 61, *b*,) is entirely black, and is sometimes found in swarms upon the potato vines in the more Northern States. There is a very similar species, the Black Blister-beetle

(*Lytta atrata*, Fabr.), from which the Black-rat Blister-beetle is distinguishable only by having four raised lines placed lengthways upon each wing-case, and by the two first joints of the antennæ being greatly dilated and lengthened in the males, as above in figure 61, *c*. The true Black Blister-beetle we have never met with, except quite late in the year, namely about the last of August or the fore part of September; usually upon the flowers of the Golden-rod, the Thistle, etc. It sometimes does injury in the potato field, especially when the development of the tubers has been retarded, but generally appears too late in the season to prove very destructive.

THE MARGINED BLISTER-BEETLE.

(*Lytta marginata*, Fabr.)

This species (fig. 62,) may be at once recognized by its general black color, and the narrow ash-gray edging to its wing-cases. It usually feeds on certain wild plants; but has been found quite abundant on potatoes in Missouri, Illinois, and elsewhere. It is a common species in the Mississippi Valley, and prefers most other varieties of the potato to the "Peach Blow." It feeds on many other plants, and especially the Kentucky Coffee-tree (*Gymnocladus*). It also attacks the Egg Plant.

Fig. 62. MARGINED BLISTER-BEETLE.

THE THREE-LINED LEAF-BEETLE.

(*Lema trilineata*, Olivier.)

The first three insects, described and figured above as infesting the potato-plant, attack it only in the larva state. The four next, namely the four Blister-beetles, attack it exclusively in the perfect state. The three that remain to be considered attack it both in the larva and in the perfect state, but go underground to pass into the pupa state; in which state—like all other beetles, without exception—they are quiescent, and eat nothing at all.

The larva of the Three-lined Leaf-beetle may be distinguished from all other insects that prey upon the potato by its habit of covering itself with its own excrement. In figure 63, *a*, this larva is shown in profile, both full and half grown, covered with the soft, greenish excrementitious matter which from time to time it discharges. Figure 63, *c*, gives a somewhat magnified view of the pupa; and figure 63, *b*, shows the last few joints of the abdomen of the larva, magnified, and viewed, not in profile, but from above. The vent of the larva, as will be seen from this last figure, is situated on the upper surface of the last joint, so that its excrement naturally falls upon its back, and by successive discharges is pushed forward towards its head, till the whole upper surface of the insect is covered with it. In other insects, which do no not indulge in this singular practice, the vent is situated either at the extreme tip of the abdomen or on its lower surface.

Fig. 63.—THREE-LINED LEAF-BEETLE.
a, Larva; *b*, End of Body; *c*, Pupa; *d*, Eggs.

There are several other larvæ, feeding upon other plants, which commonly wear cloaks of this strange material, among which may be mentioned the larvæ of certain Tortoise-beetles (*Cassida*), some of which feed on the Sweet Potato vines. (See SWEET POTATO.)

There are two broods of this species every year. The first brood of larvæ may be found on the potato vine towards the latter end of June, and the second in August. The first brood stays underground about a fortnight before it emerges in the perfect beetle state; and the second brood stays there all winter, and only emerges at the beginning of the following June. The perfect Beetle (fig. 64,) is of a pale yellow color, with three black stripes on its back, and bears a general resemblance to the common Cucumber-bug (*Diabrotica vittata*, Fabr., see fig. 27, p. 42). From this last species, however, it may be readily distinguished by the remarkable pinching in of the sides of its thorax, so as to make quite a lady-like waist there, or what naturalists call a "constriction." It is also on the average a somewhat larger insect, and differs in other less obvious respects. As in the case of the Colorado Potato-bug, the female, after coupling in the usual manner, lays her yellow eggs (fig. 63, *d*,) on the under surface of the leaves of the potato plant. The larvæ hatching from these require about the same time to develop, and when full grown, descend in the same manner into the ground, where they transform to pupæ (fig. 63, *c*,) within a small oval chamber, from which in time the perfect beetle comes forth. The remedies for the Colorado Beetle should be used for this.

Fig. 64.
THREE-LINED LEAF-BEETLE.

THE COLORADO POTATO-BEETLE.

(*Doryphora 10-lineata*, Say.)

RETROSPECTIVE.

In 1819 the United States Government fitted out an exploring expedition to the Northwest Territories under the command of Major Stephen H. Long. The zoölogist of this expedition was Mr. Thomas Say, of Philadelphia, whose name has since become so familiar to every entomologist. While on this expedition, extending through 1819 and 1820, numerous specimens of a species of beetle were found on the Upper Missouri, near the base of the Rocky Mountains, which some four years later(1824) Mr. Say described in a paper read before the Academy of Natural Sciences, Philadelphia, under the name of *Doryphora* 10-*lineata*, an insect that has since received the common name of Colorado Potato-beetle.

At the time of its discovery, neither Mr. Say nor any of his associates could have had the remotest idea that this insect would at some future day become one of the greatest pests that ever afflicted the farms and gardens of this country. Later explorers, visiting the same regions of country where Mr. Say originally found the "ten-liners," discovered it feeding on a wild species of *Solanum* (*S. rostratum*), a plant allied to and belonging to the same genus as the cultivated Potato (*Solanum tuberosum*). The pioneers on the western plains and prairies little imagined that they were in such close proximity to an insect that would soon give an immense amount of trouble, and make the cultivation of the Potato anything but a pleasant and profitable occupation. But in 1861, Mr. Thomas Murphy, of Atchison, Kansas, reported that they were so numerous in his garden that he was enabled in a very short time to gather two bushels of them. His potatoes were quickly destroyed, and the beetles then spread in all directions. Later they appeared in parts of Iowa, and subsequently passed eastward, crossing the Mississippi River, and appearing in several localities almost simultaneously within the State of Illinois. In stating that this insect passes from one locality to another, it must not be understood that it migrates, it merely spreads, enough remaining behind to keep up an abundant stock, and they are probably now no less abundant at points in the Western States than when first discovered there by Mr. Say, over sixty years ago. The sudden and enormous increase in numbers, as noted in Kansas and Iowa, was wholly due to the increase in the supply of food, for so long as this insect had to depend upon the few scattering plants of the wild Solanum, as found on the plains, its numbers were limited to a few thousands, or perhaps hundreds to the square mile; but as a single acre of potatoes will probably furnish more food than all the wild plants on a hundred acres of prairie, the sudden increase of this pest when it reached the out-lying settlements or farms of Kansas, Nebraska and Iowa, can readily be accounted for. A few years ago, their ravages in Nebraska and Kansas were severe. Since then the bugs have not caused much damage west of the Missouri.

At first the progress of the beetles eastward was at the rate of about sixty or seventy-five miles annually, but as they reached the more thickly settled regions their progress was more rapid, probably receiving some assistance from the railroads, specimens flying into the cars at some western station and escaping at another a hundred or two miles eastward, or in whatever direction the train may have been going.

NATURAL HISTORY AND TRANSFORMATIONS.

Prof. Riley was the first to make known the natural history and transformations of the Potato-beetle. They may be briefly summed up as follows: The female beetle deposits her eggs on the underside of the leaves, in clusters of a dozen, up to fifty or more. The eggs are of an orange color, and hatch in about a week after being laid, the grubs immediately commencing to feed and continuing until mature, which occurs in from fourteen to eighteen days, varying somewhat as the weather may be favorable or unfavorable. When full grown, the larvæ descend to the ground and hide under leaves or rubbish, or burrow into the soil, where they remain for ten days, then come forth in the perfect or winged form. Two to four broods are perfected during the season, according to the locality and length of the season, the last brood descending into the ground in the perfect or beetle state, and remaining in a dormant condition over winter,—reappearing as soon as the ground has become sufficiently warm to awaken

Fig. 65.—COLORADO POTATO-BEETLE (*Doryphora* 10-*lineata*).
a, a, Eggs; *b, b*, Larva in different stages; *c*, Pupa; *d, d*, Perfect Beetles of natural size; *e*, Left Wing-cover, enlarged.

them from their slumbers. The beetles at this time may usually be seen crawling about very rapidly, looking for the first shoots of the potato as it appears above ground, which they attack as though their appetite had been sharpened by a long fast.

This beetle is now too well known to need description, but it may be well to note that there is a closely allied species (*Doryphora juncta*, Germar.), often confounded with the genuine "ten-liner," although it never attacks the Potato, but feeds upon various species of wild Solanum, especially the Horse-nettle (*Solanum Carolinense*), a very common weed throughout the Middle and Southern States. Both the larva and mature insect of this Bogus Potato-beetle resemble the genuine; but upon a close examination, a very marked difference may be discovered. The most prominent distinctive characteristics observed in the nearly mature larvæ are as follows: In the true or *D.* 10-*lineata* the sides are ornamented with two rows of black dots, and the head is black; while in *juncta* there is but one row of dots, and the head is of a pale color; the first joint behind the head is reddish-brown and edged with black. The mature insects differ still more widely,

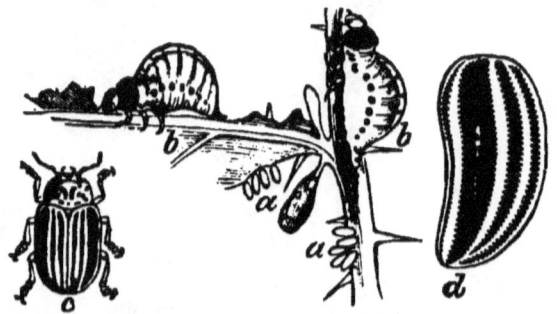

Fig. 66.—BOGUS COLORADO BEETLE (*Doryphora juncta*).
a, a, Eggs; *b, b,* Larvæ; *c,* Beetle, natural size; *d,* Left Wing-cover, enlarged.

for while 10-*lineata*, as the name indicates, has ten black stripes on its elytra, the third and fourth stripe counting from the outside, are joined behind; in *juncta*, the second and third are joined, and in a large proportion the two stripes are united the entire length, by deep brown, or black, thus forming one broad and conspicuous stripe. There are also other distinctive characters, shown in the accompanying figures, such as the arrangements of the punctures bordering the stripes on the elytra, but these are less conspicuous to the casual observer.

A few years since I tried to rear a quantity of the larvæ sent me from the South on the leaves of the Potato, but failed to carry a single specimen through to maturity on such food. The grubs will, when deprived of other and more agreeable food, attack the Potato leaves, but after eating a few moments, crawl away, and unless supplied with more of the Horse-nettle, soon die. But the genuine 10-*lineata* is not so particular in regard to its food, since the Horse-nettle and various other species of *Solanum* are just as acceptable as the Potato, and the Egg-plant (*S. melongena*) is preferred to either. On a pinch it will even feed on Jamestown-weed (*Datura*), Cabbage or Smart-weed, though it is questionable whether it could thrive for any length of time on plants belonging to other families than that of the Potato.

METHOD OF DESTROYING.

The first step or most practical method of making war upon this insect is the destruction of the few or many that come out of the ground in spring, for each female killed at this time may safely be said to represent five to ten hundred in the succeeding generation, for she will, if not prevented, lay about that number of eggs. Some persons, however, claim that it is much the best way to allow the beetles to take their own course, and then destroy the larvæ a few days later, when they have fairly commenced feeding upon the leaves, by applying some one or more of the various poisons recommended for this purpose. That either the beetles or the grubs must be destroyed in order to save the crop, is now generally admitted, and the only room for a difference of opinion is as to how it should be done. Scores of different substances have been tried for this purpose, but none have proved so effectual and economical as Paris Green and other arsenical compounds. That these poisons are dangerous to have about a place, is admitted, and so are sharp knives, reapers, and mowers, still it is not as easy to do without them as to be a little careful in using, and thereby avoid accidents. The Paris Green is destructive to the Potato-beetle in both its perfect and larval states, and one pound of the poison, mixed with twenty of pulverized plaster, or of any common kind of flour, and dusted over the leaves while wet with dew in the morning, or after a shower, will quickly cause the death of all the grubs or perfect insects feeding thereon.

A duster should be used for applying the poison, and one made of tin, with a perforated bottom, and attached to a handle four or five feet long, will be found a very convenient implement for this purpose. But the operator should be careful not to allow the compound to blow into his face, or inhale it while at work, it being only necessary for him to keep in mind that he is handling a virulent poison, and act accordingly. The Green may also be applied by mixing it with water, but as it will not dissolve, being merely suspended in the liquid, it is necessary to frequently agitate the mixture in order to prevent the poison settling to the bottom, as well as to insure its uniform distribution over the leaves. But water is a heavy material to handle, and unless one has the conveniences for applying it, the dusting process will require the least labor.

London Purple may be applied in the same way as Paris Green, and will prove equally effective, besides being much cheaper.* With most destructive beetles the larva is alone injurious, but the perfect Colorado-beetle eats as well as its larvæ.

NATURAL ENEMIES.

There are a number of other insects that aid in keeping the Colorado-beetle in check. Active among these is

---

*A more detailed history of the Colorado Beetle, as well as various forms of apparatus for distributing Paris Green and other arsenical poisons, will be found in "Potato Pests," a special treatise by C. V. Riley, of over one hundred pages.—New York : The Orange Judd Company.

the larvæ of several Lady-birds, or Lady-bugs, the perfect beetles being red, pink, or other bright color, with black spots, and generally well known by the above popular names. Their larvæ are very active and do good service in destroying both the eggs and the larvæ of the Potato-beetle. Their pupæ often resemble the larva of the Colorado-beetle, and are destroyed by mistake. Figure 67 shows one of these larvæ; the hair line gives the real size. Besides these, there are several carnivorous beetles, the Tiger-beetles, and Ground-beetles, which prey upon both the larvæ and the perfect insect. A full account of the various insects that prey upon the Colorado-

Fig. 67.
LADY-BUG LARVA.

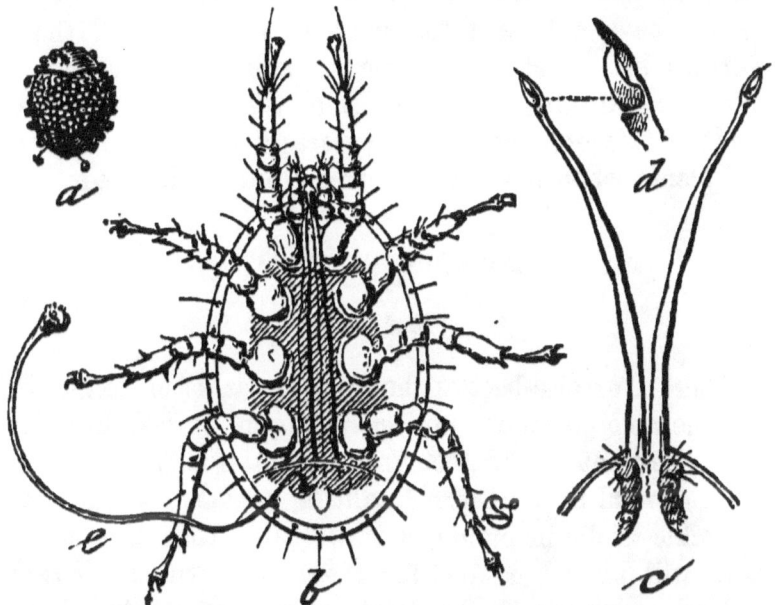

Fig. 68.—PARASITE OF COLORADO-BEETLE (*Uropoda Americana*).
a, Beetle attacked by it, natural size; *b*, Mite; *c*, penetrating or attacking organs; *d*, Claw at the end of attacking organs; *e*, Filament—all much enlarged.

beetle, is given in Prof. Riley's work, just referred to. The perfect insect is attacked by a mite which occurs in such numbers as to completely cover its victim, and it soon perishes. Figure 68, shows at *a,* the Colorado-beetle of the natural size, covered by this mite (*Uropoda Americana,* Riley), *b,* the mite greatly magnified, with a long filament which helps it to attach itself to the beetle; *c,* the penetrating organs; *d,* the claw at the end of these.

SWEET-POTATO.

The insects which attack the Sweet-potato plant are few in species, and belong almost entirely to that group of beetles popularly known as Tortoise-beetles. With the exception of the Cucumber Flea-beetle (*Haltica cucumeris,* Harr.), and a few solitary caterpillars, other insects have not been found on this plant; still these Tortoise-beetles are of themselves sufficiently numerous in individuals and species to often entirely destroy whole fields of this esculent, and they are especially severe on the plants when newly transferred from the hot-bed.

TORTOISE-BEETLES.

(*Cassidæ.*)

These Tortoise-beetles have thus far been found in considerable numbers in the Southern States, but the cultivation of the Sweet-Potato is annually becoming more general in northern localities, and as there is considerable traffic in plants, it is probable that the insect pests will spread as food for them is provided. Every one who receives Sweet-Potato plants, or "sets," from another locality, should carefully examine them before they are planted, to see that no insect is introduced with them.

These insects are almost all of a broad sub-depressed form, either oval or orbicular, with the thorax and wing-covers so thoroughly dilated at the sides into a broad and flat margin, as to forcibly recall the appearance of a turtle, whence the popular name. Many have the singular power, in a greater or less degree, of changing their color when alive, some of them shine at will with the most brilliant metallic tints.

Insects, like the higher animals, are usually cleanly in voiding their excrement, but the larvæ of several species of beetle have the peculiar habit of covering themselves with their own excrement. The larvæ of the Three-lined Leaf-beetle (*Lema trilineata*, Oliv.), which sometimes proves injurious to the potato in the East, has this habit, as do several others.

But the larvæ of the Tortoise-beetles are *par excellence* the true dung-carriers. In the instances related above, the load is carried immediately on the back, but our Tortoise-beetles are altogether more refined in their tastes, and do not allow the dung to rest on the body, but simply shade themselves with a sort of stercoraceous parasol.

The larvæ of all the species that have been observed are broad and flattened like the beetles, and have the margins of the body furnished with spines which are often barbed (fig. 75). Usually there are thirty-two of these spines, or sixteen on each side of the body. Four of these are situated on the prothorax, which forms two anterior projections beyond the common margin; four of them—the two anterior ones longer than the others—are on each of the two following thoracic segments, and each of the abdominal segments is furnished with but two. There are nine elevated spiracles each side superiorly, namely, one immediately behind the prothorax and eight on the abdominal segments. The fore part of the body is projected shield-like over the head, which is retractile and small.

The eggs from which these larvæ hatch, and which we do not recollect to have seen anywhere described, are deposited singly upon the leaves, to which they are fastened by some adhesive substance. They are of irregular angular form; flat, and somewhat narrower at one end than the other; ridged above and at the sides, but smooth and obovate below. They are usually furnished with spine-like appendages, which, however, are sometimes entirely lacking. Those of *C. aurichalcea* (fig. 69) are 0.04 inch long, and of a dull dirty-white color.

When full grown the larvæ fasten the last two or three joints of the body to the underside of a leaf, by means of

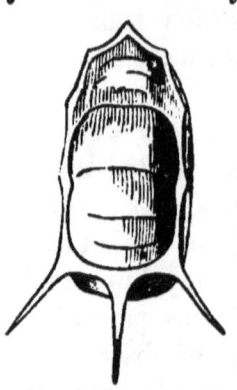

Fig. 69.—EGG OF TORTOISE-BEETLE, MAGNIFIED.

a sticky secretion, and in about two days change to pupæ. The pupa is also flat, with usually four or five broad but thin and transparent serrated leaf-like appendages on each side of the abdomen, and the prothorax, which is greatly dilated and covers the head, is furnished around the edge with smaller barbed spines. The broad leaf-like spines at the edges of the body are bent under while the transformation is being effected, but are soon afterwards stretched stiffly out with a forward slant. The pupa loses the pronged tail, but as the old larval skin is left adhering to the terminal segments the prong of dung still protects it in most cases. The legs and antennæ are not free in this, as in the pupæ of most other beetles, but are soldered together as in the chrysalis of a butterfly, and yet it has the power of raising itself up perpendicularly upon the tail end by which it is fastened. The pupa state lasts about a week.

Having thus spoken in general terms of this anomalous group of beetles, we shall now refer more particularly to a few of the species. Most of those mentioned below infest Sweet-potato both in the larva and perfect beetle states. They gnaw irregular holes, and when sufficiently numerous entirely riddle the leaves. They usually dwell on the underside of the leaves, and are found most abundant during the months of May and June. There must be several broods during the year, and the same species is often found in all stages, and of all sizes at one and the same time. In all probability they hibernate in the beetle state.

We have already proved by experiment that Paris Green—one part of the Green to two of flour—when sprinkled under the vines, will kill these insects, though not so readily as it does the Colorado Potato-beetle. Moreover, as these Tortoise-beetles usually hide on the underside of the leaves, and as the vines trail on the ground, it is very difficult to apply the powder without running some risk from its poisonous qualities. We therefore strongly recommend vigilance when the plants are first planted, and by the figures and descriptions given below our readers will be enabled to recognize and kill the few beetles which at that time make their appearance, and thus nip the evil in the bud.

THE TWO-STRIPED SWEET-POTATO BEETLE.

(*Cassida bivittata*, Say.)

This is the most common species found upon the Sweet-potato, and seems to be confined to that plant, as we have never found it on any other kind. The larva of this beetle, which is given in figure 71, 2, enlarged, and in figure 70, of natural size, is dirty-white or yellowish-white, with a more or less intense neutral-colored longitudinal line along the back, usually relieved by an extra light band each side. It differs from the larvæ of all other known species in not using its fork for merdigerous purposes. Indeed, this fork is rendered useless as a shield to the body, by being ever enveloped, after the first moult, in the cast-off prickly skins, which are kept free from excrement. Moreover, this fork is seldom held close down to the back, as in the other species, but more usually at an angle of 45° over or from the body, thus suggesting the idea of a handle.

When full fed, this larva attaches itself to the underside of the leaf, and in two days the skin bursts open on the back, and is worked down towards the tail; when the pupa, at first pale, soon acquires a dull brownish

Fig. 70.—TWO-STRIPED TORTOISE-BEETLE. Larva, natural size.

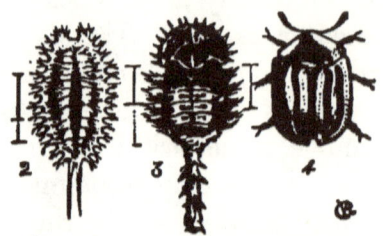

Fig. 71.—TWO-STRIPED TORTOISE-BEETLE. 2, Larva; 3, Pupa; 4, Beetle.

color, the narrow whitish tail, which still adheres posteriorly being significant of the species. (See fig. 71, 3.)

The beetle (fig. 71, 4), is of a pale yellow, striped with black, and though broader and vastly different scientifically, still bears a general resemblance to the common Striped Cucumber-beetle (*Diabrotica vittata*, Fabr.)

THE GOLDEN TORTOISE-BEETLE.

(*Cassida aurichalcea*, Fabr.)

Next to the preceding species, the Golden Tortoise-beetle is the most numerous on our Sweet-potatoes; but it does not confine its injuries to that plant, for it is found in equal abundance on the leaves of the Bitter-sweet and on the different kinds of Convolvulus or Morning Glory,

The larva (fig. 72, *a*, natural size; *b*, enlarged with the dung taken from the fork), is of a dark brown color, with a pale shade upon the back. It carries its fæcifork immediately over the back, and the excrement is arranged in a more or less regular trilobed pattern. The loaded fork still lies close to the back in the pupa, which is brown like the larva, and chiefly characterized by three dark shades on the transparent prothorax, one being in the middle and one at each side, as represented at fig. 73, *c*.

The perfect beetle (fig. 73, *d*), when seen in all its splendor, is one of the most beautiful objects that can

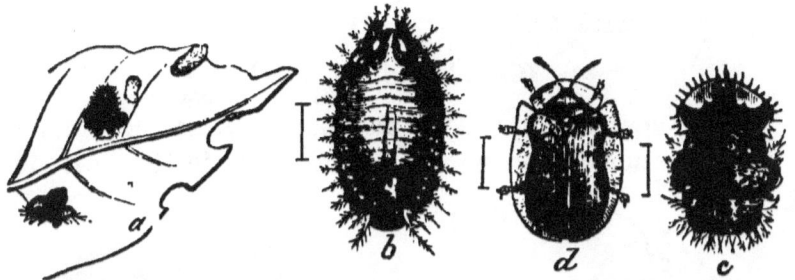

Fig. 72.—LARVA OF GOLDEN TORTOISE-BEETLE.
a, Natural size ; *b*, Enlarged.

Fig. 73.—GOLDEN TORTOISE-BEETLE.
c, Pupa ; *d*, Beetle.

well be imagined. It exactly resembles a piece of golden tinsel, and with its legs withdrawn and body lying flat to a leaf, the uninitiated would scarcely suppose it to be an insect did it not suddenly take wing when being observed. At first these beetles are of a dull deep orange color, which strongly relieves the transparent edges of the wing-coverts and helmet, and gives conspicuousness to six black spots, two (indicated in the figure) above, and two on each side. But in about a week after they have left the pupa shell, or as soon as they begin to copulate, they shine in all their splendor, and these black spots are scarcely noticed.

THE PALE-THIGHED TORTOISE-BEETLE.

(*Cassida pallida*, Herbst.)

This species can scarcely be distinguished from the preceding. It is of a somewhat broader, rounder form, and differs in lacking the black spots on the wing-coverts, and in having the thighs entirely pale yellow, while in *aurichalcea* they are black at the base. It likewise feeds upon the Sweet-potato, and its larva differs only from that of the former, in its spines being brighter and lighter colored, and in having a dull orange head, and a halo of the same color on the anterior portion of the body.

THE MOTTLED TORTOISE-BEETLE.

(*Cassida guttata*, Oliv.)

This species (fig. 74), which is the next most common of those found on the Sweet-potato in the latitude

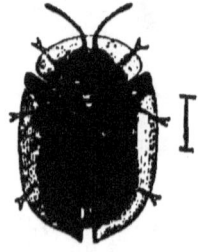

Fig. 74.—MOTTLED TORTOISE-BEETLE.

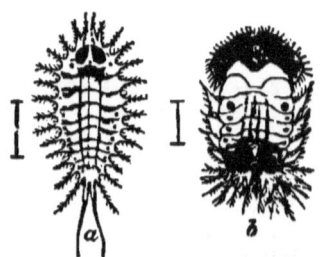

Fig. 75.—MOTTLED TORTOISE-BEETLE. *a*, Larva; *b*, Pupa.

of St. Louis, is at once distinguished from all the others here described by being usually black, with the shoulders black to the extreme edge of the transparent wing-coverts. It is a very variable species, and is frequently more or less speckled or mottled with gold, while more rarely it has a uniform golden appearance.

The larva, which is represented enlarged and with the dung removed at figure 75, *a*, is of a uniform green color, with a bluish shade along the back, which shade disappears however when the insect has fasted for a few hours. It carries its excrement in irregular broad masses, often branching as in the species next to be described. The pupa (fig. 75, *b*), is also of a uniform green color, with a conspicuous black ring around the base of the first abdominal pair of spiracles. Before changing to pupa, and previous to each moult, this larva is in the habit of removing the excrement from its fork.

THE BLACK-LEGGED TORTOISE-BEETLE.

(*Cassida nigripes*, Oliv.)

This species, which is likewise found on the Sweet-potato, is a little the largest of those we have mentioned.

Fig. 76.—BLACK-LEGGED TORTOISE BEETLE.

Fig. 77.—BLACK-LEGGED TORTOISE-BEETLE.
a, Larva of natural size; *b*, Magnified.

The beetle (fig. 76) has the power, when alive, of putting on a golden hue, but is not so brilliant as *C. aurichalcea*, from which species it is at once distinguished by its larger size, and by its black legs and three large conspicuous black spots on each wing-cover. The larva (fig. 77, *b*), is of a pale straw-color, with the spines, which are long, tipped with black; and besides a dusky shade along each side of the back, it has two dusky spots immediately beneath the head, and below these last, two larger crescent marks of the same color. The excrement is spread in a characteristic manner, extending laterally in long shreds or ramifications. (See fig. 77.) The pupa is dark brown, variegated with paler brown, while the spines around the edges are transparent and white.

TURNIP AND RUTA BAGA.

These root crops are much more generally cultivated in England than with us, and English works describe about a dozen species that are regarded as special enemies to the Turnip and Ruta Baga, or, as the latter are most commonly called, Swedes. As the cultivation of these crops becomes more general in this country, the number of destructive insects will no doubt increase.

Some of those insects that occasionally appear in great numbers, like the Fall Army Worm, and take nearly every green plant, attack the Turnip crop, though the Rocky Mountain Locust, or Grasshopper, as a rule, avoids it.

The Turnips belong to the same family of plants as the Cabbage, and several insects attack both indiscriminately. Indeed, nearly all these described under Cabbage may be looked for upon Turnips (which includes the Ruta Baga or Swedes). The Wavy-striped Flea-beetle (*Haltica striolata*), which is so destructive to young Cabbage plants, is especially fond of Turnips of all kinds in the young state, when the seedlings first break ground. This appears to be, in this country, the counterpart of the Turnip Flea-beetle of England, which is there generally called the "Turnip Fly," and is, like ours, a species of *Haltica*. If the young seedlings can be protected until they make a few rough leaves, they will usually resist these enemies, hence it has been found useful to dust them as soon as they break ground with some powder offensive to these insects. A common application is wood ashes and plaster, equal parts, the young plants to be thoroughly covered with the mixture. Air-slaked shell lime (calcined oyster-shells) is much used by market gardeners in the same manner, it is also useful as a fertilizer. Fortunately the most destructive Saw Fly and other enemies of this crop have not yet made their way to this country, but as in the exceptional season of 1881-82, large quantities of turnips were imported, it is not at all unlikely that some of the British insects may have come with them.

In the Southern States, the Harlequin Cabbage-bug (see p. 37) is very destructive to the Turnips.

Insects Injurious to the Cereal Grains, and the Grass Crops, including Clover.

In classifying insects according to the plants they injure, there is often an over-lapping. Thus the White Grub, while mentioned elsewhere, is often one of the worst enemies to the grower of grass, whether in the meadow or pasture; it also attacks the grains, as do several of the Cut-worms. When there is, as in some western localities, an invasion of the Rocky Mountain Grasshopper, scarcely any green thing escapes its attacks. We give in this division, an account of the most common enemies to the grain grower, and those which attack grass lands.

THE CHINCH-BUG.

(Blissus leucopterus, Say.)

NATURAL HISTORY OF THE CHINCH-BUG.

The food of the Chinch-bug consists of the grasses and cereals, wild and cultivated, and accounts of its injuring other plants are misleading, allied species being confounded with it. Belonging to the Half-wing Bugs (*Heteroptera*), its food is obtained by suction, so that the plants attacked are sapped of their life, and not eaten up. The mature Chinch-bug (fig. 79) is less then a fourth of an inch long; its appearance at different stages is shown in fig. 78, the hair lines indicating the natural sizes. The eggs (fig. 78, *a, b,*) are amber-colored, the young bugs vary from pale-yellow with a touch of orange to bright-red, while the pupa (*g*,) is mostly brown, the mature bug (fig. 79,) is black, with white upper wings, having two characteristic black spots upon them. A short-winged form (fig. 80,) occurs in Canada, and in the more Northern States. The species hibernates in the perfect or mature form in a state of torpor in whatever sheltered situations can be found.

The Chinch-bug is two-brooded in the Middle States, and in the more Southern States is probably three-brooded.

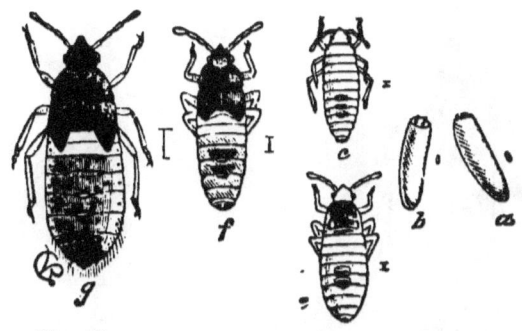

Fig. 78.—IMMATURE STAGES OF CHINCH-BUG.
a, b, Eggs ; *c* Newly-hatched Larvæ ; *f,* Same, after first Moult ; *g,* Pupa.

Such as survive the autumn, when the plants or the sap on which they feed are mostly dried up, so as to afford them little or no nourishment, pass the winter in the usual torpid state, and always in the perfect or winged form, under dead leaves, under sticks of wood, under flat stones, in moss, in bunches of old dead grass or weeds or straw, and often in corn-stalks and corn-shucks. One year I repeatedly received corn-stalks that were crowded with them, and it was difficult to find a stalk in any field that did not reveal some of them, upon stripping off the leaves.

It has long been known that the Chinch-bug deposits its eggs underground and upon the roots of plants which it infests, and that the young larvæ remain under ground for some considerable time after they hatch out, sucking the sap from the roots. If, in the spring of the year, you pull up a wheat plant in a field badly infested by this insect, you will find hundreds of the eggs attached to the roots; and at a somewhat later period the young larvæ may be found clustering upon the roots and looking like so many moving little red atoms. The egg is so small as to be scarcely visible to the naked eye, of an oval shape, about four times as long as wide, of a pale-amber-white

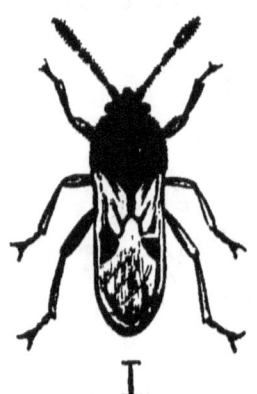

Fig. 79.—CHINCH-BUG.

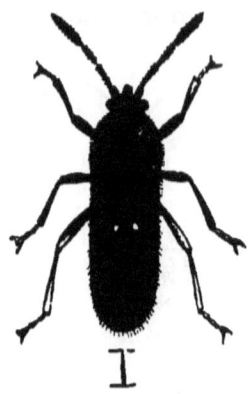

Fig. 80.—SHORT-WINGED CHINCH-BUG.

color when first laid, but subsequently assuming a reddish color from the young larva showing through the transparent shell. As the mother Chinch-bug has to work her way under ground in the spring of the year, in order to get at the roots upon which she proposes to lay her eggs, it becomes evident at once, that the looser the soil is at this time of the year the greater the facilities which are offered for the operation. Hence the great advantage of plowing land for spring grain in the preceding autumn, or, if plowed in the spring, rolling it repeatedly with a heavy roller after seeding. And the remark is frequently made by farmers, that wheat harrowed in upon old corn-ground, without any plowing at all, is far less infested by Chinch-bug than wheat put in upon land that has been plowed. There is another fact which has been repeatedly noticed by practical men. This insect cannot live and thrive and multiply in land that is sopping with water, and it generally commences its operations in early spring upon those particular parts of every field where the soil is the loosest and the driest.

There is nothing which experience has more firmly established in connection with this pest, than that heavy rains and wet seasons are destructive of it. I have witnessed the almost magical effect of a heavy and prolonged rain in a cornfield that was suffering badly. Warm, moist, or open winters are equally prejudicial to it.

The female occupies about three weeks in depositing her eggs, and, according to Dr. Shimer's estimate, she deposits about five hundred. The egg requires about two weeks to hatch, and the bug becomes full grown and acquires its wings in from forty to fifty days after hatching.

DESTRUCTIVE POWERS OF THE CHINCH-BUG.

Few persons in the more Northern States can form a just conception of the prodigious numbers and redoubtable armies in which this insect is sometimes seen in the South and South-western States, marching from one field to another. The following extracts from cotemporaneous writers I have no doubt are substantially correct, and give a clear and graphic statement of the ravages of the Chinch-bug:

There never was a better show for wheat and barley than we had here, the tenth of June, and no more paltry crop has been harvested since we were a town. Many farmers did not get their seed. In passing by a field of barley where the Chinch-bugs had been at work for a week, I found them moving in solid column across the road to a corn field on the opposite side, in such numbers that I felt afraid to ride my horse among them. The road and fences were alive with them. Some teams were at work mending the road at this spot, and the bugs covered men, horses and scrapers till they were forced to quit work for the day. The bugs took ten acres of that corn, clean to the ground, before its hardening stalks—being too much for their tools—checked their progress. Another lot passed from a wheat field adjoining my farm into a piece of corn, stopping now and then for a bite, but not long. They then crossed a meadow, thirty rods, into a sixteen-acre lot of sorghum, and swept it like a fire, though the cane was then scarce in tassel. From wheat to sorghum was at least sixty rods. Their march was governed by no discoverable law, except that they were hungry, and went where there was most to eat. Helping a neighbor harvest one of the few fortunate fields, early sown—and so lucky!—we found them moving across his premises in such numbers that they bid fair to drive out the family. House, crib, stable, well-curb, trees, garden fences—one creeping mass of stinking life. In the house as well as outside, like the lice of Egypt, they were everywhere; but in a single day they were gone.

If any Western farmer supposes that Chinch-bugs cannot be out-flanked, headed off, and conquered, they are entirely behind the times. The thing has been effectually done during the past season, by Mr. Davis, Supervisor of the town of Scott, Ogle County, Ill. This gentleman had a corn-field of a hundred acres, growing alongside of an extensive field of small grain. The bugs had finished up the latter and were preparing to attack the former, when the owner, being of an ingenious turn, hit upon a happy plan for circumventing them. He surrounded the corn with a barrier of pine boards set up edgewise and partly buried in the ground, to keep them in position. Outside of this fence deep holes were dug, about ten feet apart. The upper edge of the board was kept constantly moist with a coat of coal tar, which was renewed every day.

The bugs according to their regular tactics, advanced to the assault in solid columns, swarming by millions, and hiding the ground. They easily ascended the boards, but were unable to cross the belt of the coal tar. Sometimes they crowded upon one another so as to bridge over the barrier, but such places were immediately covered with a new coating. The invaders were in a quandary, and, in that state of mind crept backward and forward until they tumbled into the deep holes aforesaid, these were soon filled, and the swarming myriads were shovelled out of them literally by wagon loads, at the rate of thirty or forty bushels a day,—and buried up in other holes, dug for the purpose, as required. This may seem incredible to persons unacquainted with this little pest, but no one who has seen the countless myriads which cover the earth as harvest approaches, will feel inclined to dispute the statement. It is an unimpeachable fact. The process was repeated till only three or four bushels could be shovelled out of the holes, when it was abandoned. The corn was completely protected and yielded bountifully.

FALSE CHINCH-BUGS.—Some insects, with a general resemblance to the true Chinch-bug, are sometimes mistaken for that, and as they are general feeders, have given rise to reports that garden crops and others besides the grains and grasses, are attacked by the Chinch-bug. The most frequently mistaken for the true one is the False Chinch-bug (*Nysius destructor*, Riley), fig. 81, of which *b* is the pupa, and *c* the mature insect, the lines showing the real size. Its general color is grayish-brown, and that of the pupa dingy yellow. The insect is common in Missouri and Kansas. It attacks many garden vegetables, especially those of the Mustard Family (*Cruciferæ*), also the Grape-vine and Strawberry plants, to which it is especially injurious. The insect is

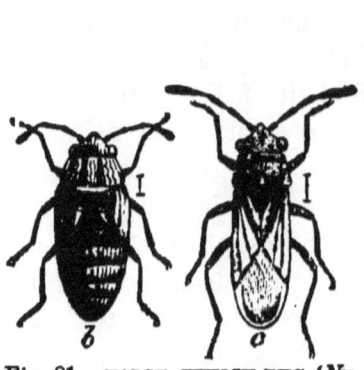

Fig. 81.—FALSE CHINCH-BUG (*Nysius destructor*, Riley).

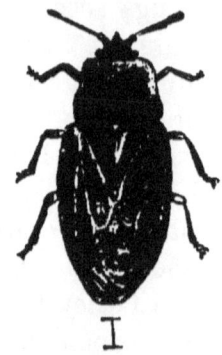

Fig. 82.—ASH-GRAY LEAF-BUG (*Piesma cinerea*, Say.)

described in full in Riley's Third Missouri Report. The Ash-gray Leaf-bug (*Piesma cinerea*), fig. 82, is often found feeding on the same plants as the Chinch-bug, and might be mistaken for that by a careless observer; a comparison of the engravings will at once show the difference.

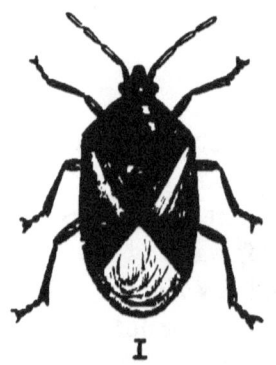

Fig. 83. INSIDIOUS FLOWER-BUG (*Anthocoris insidiosus*, Say.)

Fig. 84.—MANY-BANDED ROBBER (*Harpactor cinctus*, Fabr.)

NATURAL ENEMIES.—The Chinch-bug appears to have fewer insect enemies than other destructive insects, and this is supposed to be due to its disagreeable odor. The Insidious Flower-bug (*Anthocoris insidiosus*, Say), fig. 83, and the Many-banded Robber (*Harpactor cinctus*, Fabr.), fig. 84, are the most prominent of these, and may be recognized from the engravings. The larvæ of some of the Lady-birds, and a few others, also prey upon them more or less, but the most efficient of all are the ants, which destroy large numbers of the eggs.

REMEDIES—It has long been noticed that the Chinch Bug commences its ravages from the edges of a piece of grain, or occasionally from one or more small patches, scattered at random in the more central portions of it, and usually drier than the rest of the field. From these particular parts it subsequently spreads by degrees over the whole field, multiplying as it goes, and finally taking the entire crop unless checked up by seasonable rains. In newly broken land, where the fences are new and consequently no old stuff has had time to accumulate along them, the Chinch-bug is never heard of. These facts indicate that the mother insects must very generally pass the winter in the old dead stuff that usually gathers along fences. Hence by way of precaution, it is advisable, whenever possible, to burn up such dead stuff in the winter or early in the spring, and particularly to rake together and burn up the old corn-stalks, instead of plowing them in, or allowing them, as is often done, to lie littering about on some waste ground. It is true, agriculturally speaking, this is bad farming; but it is better to lose the manure contained in the cornstalks than to have one's crops destroyed by insects. Whenever such small infested patches in a grain field are noticed early in the season, the rest of the field may often be saved by carting dry straw on to them and burning the straw on the spot, Chinch-bugs, green wheat and all; and this will be still easier to do when the bugs start along the edge of the field. If, as frequently happens, a piece of small grain is found about harvest-time to be so badly shrunken up by the bug as not to be worth cutting, the owner ought always to set fire to it and burn it up along with its ill-savored inhabitants. Thus, not only will the insect be prevented from migrating to the adjacent corn-fields, but its future multiplication will be considerably checked.

A very simple, cheap, and easy method of prevention was recommended by Mr. Wilson Phelps, of Crete, Illinois. It may very probably be effectual when the bugs are not too numerous, and certainly can do no harm:

With twelve bushels of spring wheat, mix one bushel of winter rye, and sow in the usual manner. The rye not heading out, but spreading out close to the ground, the bugs will content themselves with eating it until the wheat is too far advanced to be injured by them. There will of course be no danger of the winter rye mixing with the spring wheat.

THE HESSIAN FLY.

(Cecidomyia destructor, Say.)

A most complete account of this insect is to be found in Bulletin No. 4, of the U. S. Entomological Commission, by Prof. A. S. Packard, of which the following is a brief abstract:

1. There are two broods of the fly, the first laying their eggs on the leaves of the young wheat from early April till the end of May, the time varying with the latitude and weather; the second brood appearing during August and September, and laying about thirty eggs on the leaves of the young winter wheat.

2. The eggs hatch in about four days after they are laid; several of the maggots or larvæ make their way down to the sheathing base of the leaf and remain between the base of the leaves and the stem, near the roots,

causing the stalks to swell and the plant to turn yellow and die. By the end of November, or from thirty to forty days after the wheat is sown, they assume the "flaxseed" state, and may, on removing the lower leaves, be found as little brown, oval, cylindrical, smooth bodies, a little smaller than grains of rice. They remain in the wheat until during warm weather in April, when the larva rapidly transforms into the pupa within its flaxseed-skin, the fly emerging from the "flaxseed" case about

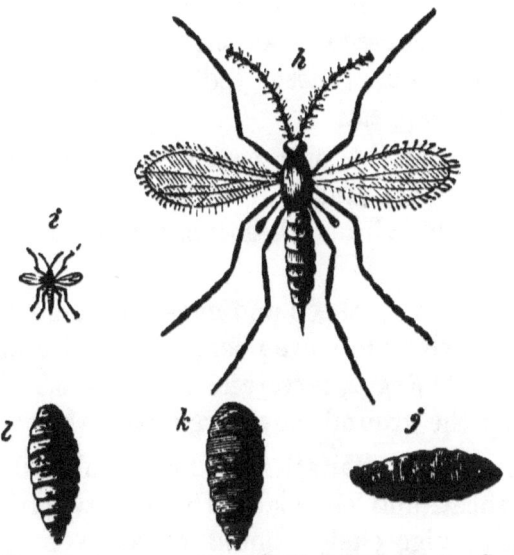

Fig. 85.—THE HESSIAN FLY (*Cecidomyia destructor*, Say.)
i, Fly of natural size; *h*, the same, magnified; *j*, *k*, Maggots, magnified; *l*, the "flaxseed" state, enlarged.

the end of April. The eggs laid by this first or spring brood of flies, soon hatch; the second brood of maggots live but a few weeks; the "flaxseed" state is soon undergone and the autumn or second brood of flies appear in August. (In some cases there may be two autumn broods, the earliest autumn brood giving rise to a third set of flies in September.) The engraving (fig. 85), shows the different states of the insect. The fly of the natural size is given at *i*, its spread of wings being only half an inch.

At *h* is the magnified insect. The body is of a dark-brown color, the wings dull smoky-brown, and the legs of a paler brown than the body. The maggots are shown, magnified in *j* and *k; l* shows the "flaxseed" state.

3. There are several destructive Ichneumon parasites of the Hessian Fly, whose combined attacks are supposed to destroy nine-tenths of all the flies hatched; of these the most important is the Chalcid four-winged fly (*Semiotellus destructor*), which infests the "flaxseed"; and the egg-parasite (*Platygaster*).

4. By sowing a part of the wheat early, and if affected by the fly, plowing and sowing the rest after September 20th, the wheat crop may in most cases be saved. It should be remembered that the first brood should be thus circumvented or destroyed in order that a second brood may not appear.

5. If the wheat be only partially affected it may be saved by fertilizers and careful cultivation; or a badly damaged field of winter wheat may thus be recuperated in the spring.

6. Pasturing with sheep and consequent close cropping of the winter wheat in November and early December may cause many of the eggs, larvæ and flaxseeds to be destroyed; also, rolling the ground may have nearly the same effect.

7. Sowing hardy varieties. The "Underhill Mediterranean" wheat, and especially the "Clawson" variety, which tillers vigorously, should be sown in preference to the slighter, less vigorous kinds, in a region much infested by the fly. The early August sown wheat might be "Diehl," the late sown "Clawson."

8. Of special remedies, the use of lime, soot, or salt, may be recommended; also raking off the stubble; but too close cutting of the wheat and burning of the stubble are of doubtful use, as this destroys the useful parasites as well as the flies.

PROBABLE GEOGRAPHICAL LIMITS OF THE HESSIAN FLY.

The question naturally arises whether this pest will ever infest the wheat regions of Western Dakota, Montana, Utah, Colorado, and the Pacific States and Territories. We believe not, though aware that such a statement may be hazardous. It was originally an inhabitant of Central and Southern Europe; it has become acclimated in the Eastern, Atlantic, and Middle States, in the Valley of the Upper St. Lawrence and in the Valley of the Mississippi River; that it can thrive in the elevated, dry Rocky Mountain plateau regions, withstand the cool nights and dry, hot atmosphere of the Far West, seems very doubtful. At least so slowly has it spread westward; so slight an amount of wheat or straw is transported, all produce of this kind going eastward, that we doubt whether during this century at least it will extend west of Kansas and Minnesota, where it has already had a foothold for several years.

Bulletin No. 4 of the Entomological Commission, by Dr. A. S. Packard, Jr., gives a full account of the Hessian Fly, which all interested should be able to procure from their representatives in Congress, as these Bulletins are published at the expense of the tax-payers.

THE WHEAT MIDGE.

(*Diplosis tritici*, Kirby.)

The Wheat Midge was formerly regarded as an insect of the same genus with the Hessian Fly, and was known as *Cecidomyia tritici*, but Entomologists now rank it in a separate genus, *Diplosis*. In general appearance the parent insect much resembles the Hessian Fly, but it deposits its eggs in the flowers of the wheat. The heads of wheat thus attacked are soon seen to shrivel, and upon examination there will be found numerous legless maggots, about one-twelfth of an inch long, and of an orange color, among the forming grain, which are popularly known as midges, a portion of the larvæ or midges go into the ground and pupate, while others are harvested with the grain. Some parasitic insects help reduce the numbers of the midge, and so far as is known, deep plowing, to turn those which have entered the ground so deep that they cannot make their way to the surface, and the burning of the refuse in the cleaning of the grain, are the only artificial helps suggested.

THE JOINT-WORM.

(*Isosoma hordei*, Harris.)

In certain years and in particular States the crops of wheat, of barley, or of rye are greatly injured by a minute maggot, popularly known as the "Joint-worm." This maggot is but little more than one-eighth of an inch long, and of a pale-yellow color with the exception of the jaws, which are dark-brown. It inhabits a little cell, which is situated in the internal substance of the stem of the affected plant, usually a short distance above the first or second knot from the root, the outer surface of the stem being elevated in a corresponding elongate blister-like swelling; and when, as is generally the case, from three to ten of these cells lie close together in the same spot, the whole forms a woody enlargement, honey-combed by cells, and is in reality a many-celled gall. In figure 86, *a*, will be seen a sketch of one of these galls, the little pin-holes being the orifices through which the flies produced from the joint-worms have escaped. At first sight, these knotty swellings of the stem are apt to elude observation, because, being almost always situated just above the joint or knot on that stem—whence comes the popular name "Joint-worms"—they are enwrapped and hidden by the sheath of the blade; but on stripping off the sheath, as is supposed to have been done in the engraving, they become at once very conspicuous objects. We have observed that the "internodes," as botanists call them, or the spaces between the knots, in infested straws are always much contracted in length, none out of a lot of over fifty specimens examined by us exceeding six inches in length, and many being reduced to only one and a half inch. There were only three straws in this lot of over fifty straws, where two Joint-worm galls were

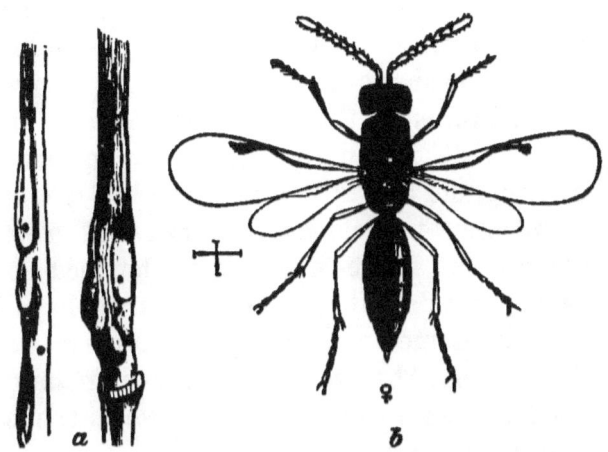

Fig. 86.—THE JOINT-WORM (*Isosoma hordei*, Harris).
a, Galls at joints; *b*, Female Fly, enlarged, the lines showing the real size.

found in the same straw; and in all those three cases they were found in two adjoining internodes. In a very few instances the galls were situated in the middle of the internode, or even close to the upper knot, instead of being situated as usual above the lower knot.

AMOUNT OF DAMAGE DONE BY THE JOINT-WORM.

The damage occasioned by the Joint-worm is, in certain seasons and in certain localities, ruinously great. In the year 1851, through a large part of Virginia, according to the Editor of the "Southern Planter," "many crops of wheat were hardly worth cutting on account of its attacks, and all that we have seen or heard of, except one, were badly hurt by it." It first began to be observed in that State in 1848, and in subsequent years it increased gradually in numbers. According to Prof. Cabell, of the University of Virginia, the loss occasioned by this insect often amounted to one-third of the average crop, and sometimes much greater; and in 1851 "some farmers did not reap as much as they sowed." In 1860 the rye crop was considerably injured by this little pest in Lycoming Co., Pennsylvania; and according to Mr. Norton, the species is very common upon rye "in Connecticut and probably the other New England States." As long ago as 1829, it had been noticed in various parts of the New England States to attack the barley, causing it in some places "to yield only a very small crop, and on some farms not much more than the seed sown;" although since that date it does not appear to have been materially troublesome in that region. But in Central New York, formerly the great barley-growing district of America, it has been ruinously destructive to the barley from 1850 until the present.

It is a curious fact that—so far as can be at present ascertained—this destructive insect does not appear to have reached the Valley of the Mississippi. At all events, no complaints from the West of any such attacks as those described above, either upon wheat, rye, or barley, have hitherto been make public. It is very possible, however, that the Joint-worm may have been confounded in the West with the Hessian Fly (*Cecidomyia destructor*, Say), the larva of which infests the same part of the wheat plant, namely, the space immediately above one of the lowermost knots in the straw. But this last may be distinguished from the Joint-worm by living in the open space between the stem and the sheath of the blade, although it occasionally imbeds itself pretty deeply in the external surface of the stem; whereas, the true Joint-worm always inhabits a smooth egg-shaped cell in the internal substance of that stem.

NATURAL HISTORY OF THE JOINT-WORM.

The mode in which the Joint-worm produces its destructive effects upon small grain, may be readily explained. Not only is the sap extracted on its course to the ear, in order to form the abnormal woody enlargement or gall, in which the larvæ are imbedded, each in his own private and peculiar cell, but a very large supply of sap must be wasted in feeding the larvæ themselves. Hence the ear that would otherwise be fully developed becomes more or less shriveled; although we are told that, in the case of barley more particularly, the plant tillers out laterally, so as partially to supply the loss of the main crop of ears. A similar phenomenon occurs with almost all galls that grow upon a slender stem or twig, that is, the stem or twig is more or less killed or blasted thereby; but when a twig is quite large, this result often fails to be developed.

The Joint-worm Fly (fig. 86, *b*,) makes its appearance in the North in the forepart and middle of June, and in southern latitudes in the middle of May. After coupling in the usual manner, the female Joint-worm Fly proceeds to lay her eggs in the stems of the growing grain. The following excellent account of this operation, from the pen of Mr. Pettit, is from the "Canada Farmer":

"About the eighth of June, the perfect insects begin to make their way out of the galls. * * * * * I watched the growing barley, and on the tenth found them actively at work ovipositing in the then healthy stalks of the plant. Before commencing operations they walk leisurely up one side of the plant as far as the last leaf, and then down the other, apparently to make sure that it has not already been oviposited in. Head downward, they then begin by bending the abdomen downward, and placing the tip of the ovipositor on the straw at right angles with the body, when the abdomen resumes its natural position, and the ovipositor is gradually worked into the plant to its full extent. With the aid of a good lens, and by pulling up the plants on which they were at work (which did not appear to disconcert them in the least), I could view the whole operation, which, in some cases, was accomplished in a few minutes, and in others was the work of an hour or two. When a puncture was completed, they usually backed up a little and viewed it for a few seconds, and then apparently satisfied, moved to one side and another began."

Very shortly after this time, the egg must hatch out. For, upon July third, we examined a large lot of the green barley-galls, which had been obligingly forwarded to us by Mr. Pettit, and found the larva of the Jointworm Fly almost half-grown, that is from 0.004 to 0.006-inch long, and about five times as long as wide.

By the beginning of September, the infested grain having ripened long before this period, the galls are already dry and hard, and the larvæ contained in them full grown, measuring now about 0.13-inch in length. The great majority of these larvæ are destined to remain in that state, enclosed in their little cells, until the succeeding spring; but—as happens with many different insects— a small percentage of them seem to pass into the pupa, and thence into the perfect state, the same summer that the eggs are deposited. For, out of a lot of one hundred and twenty-four barley-galls, received September 10th from Mr. Pettit of Upper Canada, thirty-nine galls, on very nearly one-third part, were already bored with the same kind of small round holes as are made in the succeeding spring by the escaping Jointworm Flies, some galls containing six such holes, but most of them about three. It is true that we are not personally cognizant of the fact, that these holes are bored by the same Joint-worm Fly, that escapes from similar holes in such profuse abundance in the following June; but Prof. Cabell, of Virginia, stated to Dr. Harris with reference to the wheat-inhabiting Joint-worm, that he had known a few flies to leave the straw the first year, but in each instance the fly which came forth thus was the true Joint-worm Fly. As already shown, the flies that emerged from these Canada galls in the succeeding summer, came out from June 9th to June 16th and subsequently.—(American Entomologist.)

ARMY WORMS.

The name Army Worm is somewhat loosely applied to several different insects that have the habit of congregating in considerable numbers, or in moving from place to place in large bodies. In some localities in Western New York, the name is given to the Tent Caterpillar of the Forest (*Clisiocampa sylvatica*, Harr.), described under FRUIT TREES.

In some of the Southern States, the Cotton Worm (*Aletia argillacea*, Hubn.), is called "Army Worm," and more frequently the "Cotton Army Worm," an insect most exhaustively treated of by Prof. Riley in Bulletin No. 3, of the U. S. Entomological Commission.

Still another insect, common in the Southern States, (*Laphrygma frugiperda*, Sm. and Abb.), which sometimes attacks cotton, has been called "Army Worm." Its proper name is "Southern Grass Worm," and it prefers grasses and weeds to cotton and other crops. To distinguish the true Army Worm from all others to which the name has been given, it may be called:

THE NORTHERN ARMY WORM.

(Leucania unipuncta, Haw.)

This insect has from time to time made its appearance in destructive numbers. Its earliest recorded appearance in the Eastern States, was in 1743. The years 1770, 1817, and 1861, are those in which it is reported to have been especially troublesome in the East; in 1861 it was destructive from New England to Kansas; in 1875, it visited a large part of Missouri, and in 1880 was especially destructive on Long Island. Prof. C. V. Riley was the first to give the full history of this insect, in his Reports on the Insects of Missouri, and in the Walker Prize Essay of the Boston Natural History Society, for 1877; from these the following is condensed.

DESCRIPTION OF THE INSECT.

The worm when full grown is dingy black in color, striped as in figure 87, with a broad dusky stripe along the back, divided along the middle by a more or less distinct and irregular pale line, and bordered beneath by a narrow black line; then a narrow white line; then a yellowish stripe; then a narrow, indistinct white line; then another dusky stripe; again a narrow white line; then a yellow stripe, and, finally, again a faint white line: the underside or venter is obscure green.

The chrysalis (fig. 88) is mahogany-brown in color. The moth (fig. 89) is of a fawn color, with a white speck near the center of the front wings and a dusky, oblique line running inwardly from their tips.

The eggs are laid in the spring of the year so far as we know, and not in the fall as was formerly supposed. They are thrust, by means of an ovipositor, which is admirably adapted for this purpose, in between the folded sides of a grass blade and glued along the grooves with a white, glistening, and adhesive fluid, which not only fastens them together but draws the two sides of the grass blade close around them so as to pretty effectually hide them. The female performs this operation at night, and is extremely active at the time, laying her eggs with great rapidity, so that the ovaries are soon emptied. Each individual egg is glistening white at first, but becomes dull yellowish toward maturity. The female prefers a dry blade to a green one, and is espe-

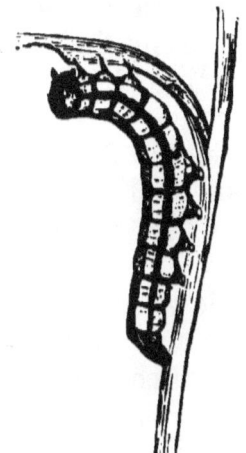

Fig. 88.—CHRYSALIS.

Fig. 87.—ARMY WORM. Fig. 89.—MOTH OF ARMY WORM.

cially prone to oviposit in places where there is a thick matting of coarse, last year's grass. The young worm hatches in about ten days, and up to the last moult has all the habits of an ordinary Cut-Worm, the colors being much paler than when full grown, and the worm hiding during the day at the base of the grasses. When not excessively numerous they retain this their normal Cut-Worm habit, and only when they become excessively multiplied do they acquire the marching and migrating habits.

REMEDIES.—Experience has established the fact that burning a meadow, or prairie, or field of stubble, in winter or spring, usually prevents the worms from originating in such meadow or field. Such burning destroys the previous year's stalks and blades, and, as a consequence of what we have already stated, the nidi which the female moth prefers. Burning as a preventive, however, loses much of its practical importance unless it is pursued annually, because of the irregularity in the appearance of the Worm in injurious numbers. Judicious ditching, *i. e.*, a ditch with the side toward the field to be protected perpendicular or sloping under, will protect a field from invasion from some other infested region when the worms are marching. When they are collected in the ditch they may be destroyed either by covering them with earth that is pressed upon them, by burning straw over them or by pouring a little coal oil in the ditch. A single plow furrow, six or eight inches deep and kept friable by dragging brush in it, has also been known to head them off.

From experiments which we have made we are satisfied that where fence-lumber can be easily obtained it may be used to advantage as a substitute for the ditch or trench, by being secured on edge and then smeared with kerosene or coal tar, the latter being more particularly useful along the upper edge. By means of laths and a few nails the boards may be so secured that they will slightly slope away from the field to be protected. Such a barrier will prove effectual where the worms are not too persistent or numerous. Where they are excessively abundant they will need to be watched and occasionally dosed with kerosene to prevent their piling up even with the top of the board and thus bridging the barrier. The lumber is not injured for other purposes subsequently. In the invasion of Long Island in 1880, but two methods were found successful in checking the march of the Army Worm. Trenches were made by plowing, and in these were distributed freshly cut Red-top grass, a favorite food with them, and the grass was sprinkled with a mixture of Paris Green or London Purple in water, the same that is used for the Colorado Potato-beetle. So long as the grass remained fresh, the worms were destroyed by millions. Trenches by themselves were of little use, but if pits are made at every rod or so in the trench, about a foot square, and two feet deep with clean straight sides, the worms, in seeking a place to escape from the trench, will fall into these pits in great numbers. When one pit is nearly full of worms, others may be dug, using the earth to bury those already in the pits. The trenches should be dressed with the spade, after the plow, to make sure of straight smooth sides.

SUMMARY.

The following summary of the natural history of the Worm is from the 9th Missouri Report:

"The insect is with us every year. In ordinary seasons, when it is not excessively numerous, it is seldom noticed. 1st, because the moths are low, swift flyers, and nocturnal in habit; 2nd, because the worms, when young, have protective coloring, and, when mature, hide during the day at the base of grasses. In years of great abundance the worms are generally unnoticed during early life, and attract attention only when, from crowding too much on each other, or from having exhausted the food supply in the fields in which they hatched, they are forced from necessity to migrate to fresh pastures in great bodies. The earliest attain full growth and commence to travel in armies, to devastate our fields, and to attract attention, about the time that winter wheat is in the milk—this period being two months later in Maine than in Southern Missouri; and they soon afterwards descend into the ground, and thus suddenly disappear, to issue again in two or three weeks as moths. In the latitude of St. Louis, the bulk of these moths lay eggs, from which are produced a second generation of worms, which become moths again late in July or early in August. Exceptionally a third generation of worms may be produced from these. Further north there is but one generation annually. The moths hibernate, and oviposit soon after vegetation starts in spring. The chrysalides may also hibernate, and probably do so to a large extent in the more Northern States. The eggs are inserted between the sheath and stalk, or secreted in the folds of a blade; and mature and perennial grasses are preferred for this purpose. The worms abound in wet springs preceded by one or more very dry years. They are preyed upon by numerous enemies, which so effectually check their increase, whenever they unusually abound, that the second brood, when it occurs, is seldom noticed; and two great Army Worm years have never followed each other, and are not likely to do so."

THE WHEAT-HEAD ARMY WORM.

(*Leucania albilinia*, Guen).

There has of late years appeared, first in Pennsylvania and Maryland, and later in Kansas and Missouri, an insect in many respects like the true Army Worm, but which has shown a peculiar tendency to feed upon the heads of wheat and other small grains. When newly hatched, this differs from the true Army Worm by its black head and later by having five instead of seven pale lines, and six instead of eight dark ones. The habit of feeding upon the grain becomes fixed only when the worms are half grown, as before that they attack the leaves, grass, etc. Several parasitic insects diminish its numbers, and it has been suggested that the worms could be greatly diminished by setting traps to attract the moths by means of lights to poisoned sweet liquids.

CLOVER.

In an invasion of the Rocky Mountain Locust or Grasshopper, the Clover suffers with most other green things, but the generally voracious Army Worm, while it occasionally nibbles at it, usually passes to more acceptable plants. There are a few caterpillars of moths now and then found upon Clover, but are regarded mainly as accidental. Within a few years, it has been discovered that

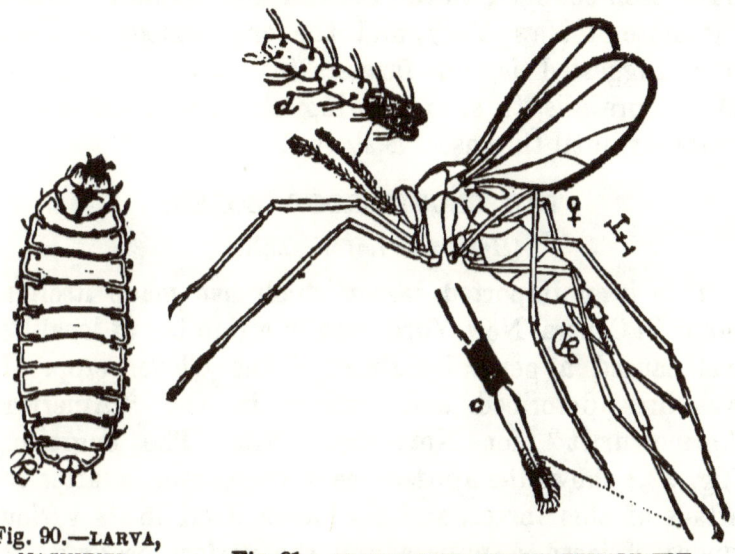

Fig. 90.—LARVA, MAGNIFIED.

Fig. 91.—CLOVER-SEED MIDGE, FEMALE FLY.
c, Ovipositor; d, Joints of Feeler, enlarged.

this important plant has two enemies, both of such a destructive character as to demand attention, one, a midge that destroys the seed, the other a borer that injures the root.

THE CLOVER-SEED MIDGE.
(*Cccidomyia trifolii*, Riley.)

This insect was described and figured in the "American Agriculturist" for July 1879. The larva (fig. 90, much magnified), is of a bright orange-red color, lives within the Clover-heads, and by exhausting them of their juices, causes the seed to shrivel and become worthless. When they have made their growth, these Midges either enter the ground, or hide under rubbish on the surface, and form a tough silken cocoon, with particles of earth adhering. Some of the flies appear in September, and others not until the following spring. Figure 91 gives a highly magnified view of the female fly and its details. Thus far this insect's ravages have been confined to the Central and Western parts of the State of New York, and the only remedy that has been suggested is, for farmers in localities where the Midge prevails, to stop growing clover-seed for several years, or until the insect is starved out.

THE CLOVER-ROOT BORER.
(*Hylesinus trifolii*, Miller.)

This is an imported insect which has made itself at home in Central New York, and in a number of localities has caused a general failure of the Clover crop. It was first described and figured in the "American Agriculturist" for November 1879. The engraving (fig. 92,) shows the appearance of the Clover, *a*, after the attack of this insect, and the insect itself in its various stages of larva, *b*, pupa *c*, and the perfect beetle *d*. It passes the winter in either of these three states, and in early spring the insects issue and pair. The female then instinctively bores into the crown of the root, eating a pretty large cavity, wherein she deposits from four to six pale, whitish, elliptical eggs. These hatch in about a week, and the young larvæ at first feed in the cavity made by the parent. After a few days, however, they begin to burrow downward, extending to the different branches of the root. The galleries made in burrowing run pretty regularly along the axis of the roots, as shown in the engraving, and are filled with brown excrement. The pupa is formed in a smooth cavity, generally at the end of one of these burrows, and may be found in small numbers as early as September.

It is the custom in Western New York to sow the Clover in spring on ground already sown to fall wheat. This is generally done while snow is yet on the ground, or while the frost is disappearing. The Clover is allowed to go to seed in the fall, and usually produces but little. During the second year one crop of hay and a crop of seed are obtained. It is during the second year the injury of the Root-borer is most observed.

One observer reports that this insect has attacked all the clover in portions of Genesee County. I examined clover in some half a dozen fields during a ride of ten miles, and found every plant I pulled up was more or less injured. While most of the plants are yet alive, they are of little value for hay, seed, or pasture. The only remedy thus far suggested is, to plow under all the clover found to be infested in the spring of the second year. Some parasites are known to prey on this insect, which may diminish it.

Fig. 92.—CLOVER-ROOT BORER (*Hylesinus trifolii*, Miller.)
a, Injured stem and root; *b*, Larva; *c*, Pupa; *d*, Beetle, enlarged.

THE CLOVER-WORM.
(*Asopia costalis*, Fabr.)

This insect, like the preceding introduced from Europe, has been occasionally noticed for the last twenty years, and now, in some localities, from New England to Michigan and Illinois, it often occurs in troublesome numbers. It attacks the clover in the mow or stack, webbing the stems together with multitudes of silken threads, among which is such an abundance of black excrement as to unfit the clover for feeding to animals. The white cocoons are present in such numbers, that one, without close examination, would pronounce the hay to be mouldy.

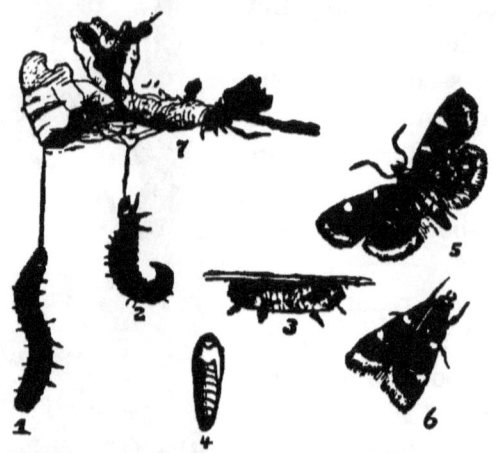

Fig. 93.—CLOVER-WORM (*Asopia costalis*, Fabr.)
1 and 2, Larva; 3, Cocoon, 4, Pupa; 5 and 6, the Moth, 7, the Web.

The insects are usually found at the bottom of the stack. Figure 93 shows the insect in its various stages, 1 and 2 represent the larva, 3 the cocoon, 4 the pupa, 5 and 6 the moth, and 7 the white web in which the worm for the most part lives. The moth is one of our prettiest species, being of a reddish-brown color with golden-yellow markings and fringe to its wings. It is suggested as a preventive, that hay containing clover should not be stacked twice in the same place, and that the stack should be placed upon log or other foundations, that will allow of thorough ventilation from below.

Insects Injurious to Fruit Trees.

APPLE-TREE BORERS.

THE ROUND-HEADED APPLE-TREE BORER.

(*Saperda bivittata*, Say.)

It is an admitted fact that apple trees on the ridges are shorter lived than those grown on our lower lands. Hitherto no particular reason has been given for this occurrence, but it appears to be mainly attributed to the workings of the borer now under consideration. It has been invariably found more plentiful in trees growing on high land than in those on low land, and worse in plowed orchards than on those which are seeded down to grass. Fifty years ago, large, thrifty, long-lived trees were exceedingly common, and were obtained with comparatively little effort on the part of our ancestors. They had not the vast army of insect enemies to contend with, which at the present day makes successful fruit growing difficult. This Apple-Tree Borer was entirely unknown until Thomas Say described it in the year 1834; and, according to Dr. Fitch, it was not until the year following that its destructive character became known in the vicinity of Albany, N. Y., for the first time. Yet it is a native American insect, and has for ages inhabited our indigenous Crab-apple trees from which trees Mr. A. Bolter took numerous specimens, in the vicinity of Chicago, ten years ago. It also attacks the Quince, Mountain-ash, Hawthorn, Pear, and the June-berry. Few persons are aware to what an alarming extent this insect is infesting the orchards in various localities. A tree becomes unhealthy and eventually dwindles and dies, often without the owner having the least suspicion of the true cause—the gnawing worm within.

At figure 94 this borer is represented in its three stages of larva (*a*), pupa (*b*), and perfect beetle (*c*). The beetle may be known by the popular name of the Two-striped Saperda, while its larva is best known by the name of the Round-headed Apple-Tree Borer, in contradistinction to the flat-headed species next treated of.

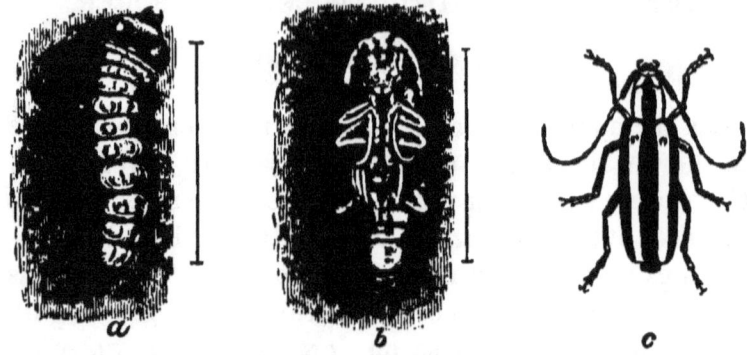

Fig. 94.—ROUND-HEADED APPLE-TREE BORER (*Saperda bivittata*, Say.)
a, Larva; *b*, Pupa; *c*, Beetle.

The average length of the larva, when full grown, is about one inch, and the width of the first segment is not quite one-fourth of an inch. Its color is light-yellow, with a tawny-yellow spot of a more horny consistency on the first segment, which, under a lens, is found to be formed of a mass of dark-brown spots. The head is chestnut-brown, polished and horny, and the jaws are deep-black. The pupa is of rather lighter color than the larva, and has transverse rows of minute teeth on the back, and a few at the extremity of the body; the perfect beetle has two longitudinal white stripes between three of a light cinnamon-brown color. The Two-striped Saperda makes its appearance in the beetle state during the months of May and June, and is seldom seen by any but the entomologist who makes a point of hunting for it, as it remains quietly hidden by day and flies and moves only by night. The female deposits her eggs during the month of June, mostly at the foot of the tree, and the young worms hatch and commence boring into the bark within a fortnight afterwards. These young worms differ in no essential from the full grown specimens, except in the very minute size; and they invariably live for the first year of their lives on the sap-wood and inner bark, excavating shallow, flat cavities which are found stuffed full of their saw-dust like castings. The hole by which the newly hatched worm penetrates is so very minute that it frequently fills up, though not before a few grains of castings have fallen from it, but the presence of the worms may be generally detected, especially in young trees, from the bark, under which they lie, becoming darkened, and sufficiently dry and dead to contract and form cracks. Through these cracks, some of the castings of the worm generally protrude, and fall to the ground in a little heap, and this occurs more especially in the spring of the year, when, with the rising sap and frequent rains, such castings become swollen and augment in bulk. Some have supposed that the worm makes these holes to push out its own excrement, and that it is forced to do this to make room for itself; but, though it may sometimes gnaw a hole for this purpose, such an instance has never come to our knowledge, and that it is necessary to the life of the worm is simply a delusion, for there are hundreds of boring insects who never have recourse to such a procedure, and this one is frequently found below the ground, where it cannot possibly thus get rid of its castings. It is currently supposed that this borer penetrates into the heart wood of the tree after the first year of its existence, whereas the Flat-headed borer is supposed to remain for the most part immediately under the bark; but on these points no rule can be given, for the Flat-headed species also frequently penetrates into the solid heart wood, while the one under consideration is often found in a full grown state just under the inner bark, or in the sap wood. The usual course of its life however runs as follows:

As winter approaches, the young borer descends as near the ground as its burrow will allow, and doubtless remains inactive until the following spring. On approach of the second winter it is about one-half grown, and still living on the sap-wood; and it is at this time that these borers do the most damage, for where there are four or six in a single tree, they almost completely girdle it. During the next summer, when the worm has become about three-fourths grown, it generally commences to cut a cylindrical passage upward into the solid wood, and before it has finished its larval growth, it invariably extends this passage right to the bark, sometimes cutting entirely through a tree to the opposite side from which it commenced; sometimes turning back at different angles. It then stuffs the upper end of the passage with sawdust-like powder, and the lower part with curly fibres of wood, after which it rests from its labors. It thus finishes its gnawing work during the commencement of the third winter, but remains motionless in the larval state until the following spring, when it casts off its skin once more and becomes a pupa. After resting three weeks in the pupa state, it appears as a beetle, with all its members and parts at first soft and weak. These gradually harden, and in a fortnight more it cuts its way through its sawdust-like castings, and issues from the tree through a perfectly round hole. Thus it is in the tree a few days less than three years, and not merely two years as Dr. Fitch suggests.

REMEDIES.—From this brief sketch of the Round-headed borer, it becomes apparent that plugging the holes to keep him in, is on a par with locking the stable door to keep the horse in, after he is stolen; even supposing there were any philosophy in the plugging system, which there is not; the round smooth holes are infallible indication that the borer has left, while the plugging up of any other holes or cracks where the castings are seen, will not affect the intruder. This insect probably has some natural enemies, belonging to its own great class, and some wood-peckers doubtless seek it out from its retreat and devour it; but its enemies are certainly not sufficiently under our control, and to grow healthy apple trees, we have to fight it artificially. Here again prevention will prove better than cure, and a stitch in time will not only save nine, but fully ninety-nine.

Experiments have amply proved that alkaline washes are repulsive to this insect, and the female beetle will not lay her eggs upon trees protected by such washes. Keep the base of every tree in the orchard free from weeds and trash, and apply soap to them during the month of May, and they will not probably be troubled with borers. For this purpose soft-soap or common bar soap can be used. The last is perhaps the most convenient, and the newer and softer it is the better. Home-made soft-soap, such as is prepared on many farms from ley of wood-ashes, usually contains an access of alkali, and when thinned with water, so that it will work with a brush is excellent. This borer confines himself almost entirely to the base of the tree, though very rarely it is found in the crotch. It is therefore only necessary in soaping, to rub over the lower part of the trunk and the crotch, but is a very good plan to lay a piece of hard soap in the principal crotch, so that it may be washed down by the rains. In case these precautions have not been taken, and the borer is already at work, many of them may be killed by cutting through the bark at the upper end of their burrows, and gradually pouring hot water into the cuts, so that it will soak through the castings, and penetrate to the insect. But even where the soap preventive is used in the month of May, it is always advisable to examine the trees in the fall, at which time the young worms that hatched through the summer may be generally detected, and easily cut out without injury to the tree. Particular attention should also be paid to any tree that has been injured or sun-scalded, as such trees are most liable to be attacked.

THE FLAT-HEADED APPLE-TREE BORER.

(*Chrysobothris femorata*, Fabr.)

This borer which is presented in the larva state at figure 95, *a*, may at once be recognized by it anterior ends being enormously enlarged and flattened. It is paler than the preceding, and makes an entirely different burrow. In consequence of its immensely broad and flattened head, it bores a hole of an oval shape, and twice as wide as high. It never acquires much more than half the size of the other species, and is almost always found with its tail curled completely round towards the head. It lives but one year in the tree, and produces the beetle represented at figure 95, *d*, which is of a greenish-black color, with brassy lines and spots above, the underside appearing like burnished copper. This beetle flies by day instead of by night, and may often be found on different trees basking in the sunshine. It attacks not only the Apple, but the Peach, also the Soft Maple, Oak, and is said to attack a variety of other forest trees; though, since the larvæ of the family (*Buprestidæ*), to which it belongs, all bear a striking resemblance to each

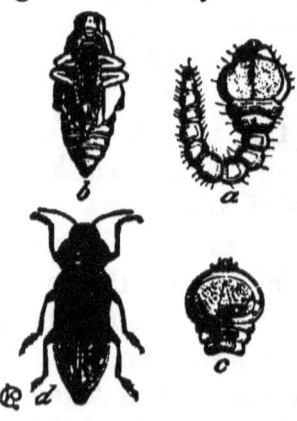

Fig. 95.—FLAT-HEADED APPLE-TREE BORER (*Chrysobothris femorata*, Fabr.) *a*, Larva; *b*, Pupa; *c*, upper joints of Larva seen from beneath; *d*, Beetle.

other, it is possible that this particular species has been accused of more than it deserves. It is, however, but far too common in the Valley of the Mississippi, and along the Iron Mountain and Pacific railroads; it is even more common than the preceding species.

Mr. G. Paul, of Eureka, states that it has killed fifty apple trees for him, and Mr. Votaw, and many others in that neighborhood have suffered from it in like manner. It is also seriously affecting the soft maples by riddling them through and through, though it confines itself for the most part to the inner bark, causing peculiar black scars and holes in the trunk. Unless its destructive work is soon checked, it bids fair to impair the value of this tree for shade and ornamental purposes, as effectually as the Locust-borers have done with the Locust trees.

REMEDIES.—Dr. Fitch found that this borer was attacked by the larvæ of some parasitic fly belonging probably to the *Chalcis* family, but it is greatly to be feared that this parasite is as yet unknown in the West. At all events this Flat-headed borer is far more common with our Eastern brethren. As this beetle makes its appearance during the months of May and June, and as the eggs are deposited on the trunk of the tree, as with the preceding species, the same method of cutting them out, or scalding them can be applied in the one case as in the other; while the soap preventive is proved to be equally effectual with this species as with the other. It must, however, be applied more generally over the tree, as they attack all parts of the trunk, and even the larger limbs.

THE APPLE-TWIG BORER.

(Amphicerus bicaudatus, Say.)

The Apple-twig borer is a modest looking dark-brown insect, the thorax rounded and rough-punctured, especially in front where it is produced into two little horns, and covered with small rasp-like prominences. The wing-covers are also rough-punctured, and while in the female (fig. 96, *a*), they have but a slight keel-like elevation at the hind end, they are furnished in the male (fig. 96, *b*), with two little horns, from which characteristic the specific name (two-tailed) is derived.

The holes made in the twigs, generally have their entrance just above a bud or fork as at figure 97, *c*. This insect is not known to bore more than an inch and a-half into the twig (fig. 97, *d*), and the holes are generally made downwards, and in the wood of the previous year's growth, though they are sometimes exceptionally bored upward and in three-year old wood. The beetles seem to prefer some particular varieties, such as Benoni and

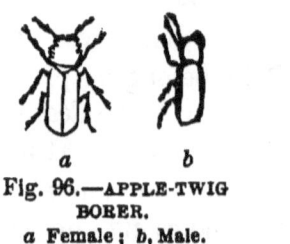

Fig. 96.—APPLE-TWIG BORER.
a Female; *b*, Male.

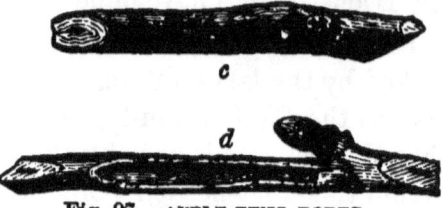

Fig. 97.—APPLE-TWIG BORER.
c, Puncture; *d*, Interior of Stem.

Red June, to other varieties of the Apple, and though they likewise occur in Pear and Peach stems, and in the Grape, they have not been found in those of the Crab-apple.

Both the male and female beetles bore these holes, and may always be found in them, head downwards, during the winter and spring months. The holes are made for food and protection, and not for breeding purposes. Indeed, common as this insect is, its preparatory stages are entirely unknown, and whoever will ascertain its larval history, will confer a favor on the community.

The bored twigs almost always break off by the wind, or else the hole catches the water in spring and causes an unsound place in the tree. If the twig does not break off, it withers and the leaves turn brown. The only way to counteract the injuries committed by this beetle, is to prune the infested twigs, whenever found, and take great care to burn them with their contents. It is in the nursery that most damage is done by this insect, as it is seldom numerous enough in an orchard of large trees to more than cause what the philosophic orchardist has termed "a good summer pruning."

BARK-LICE.

The Bark-lice belong to the Order *Hemiptera,* in which they form the group or family, *Coccideæ,* so named from the genus *Coccus,* one species of which is the remarkable Cochineal Insect. Several of these insects are very injurious to the Orange trees and others of that Family; one infests the Osage Orange, while at least two attack our orchard trees, especially the Apple, though the Pear, Quince, etc., are often infested by them.

HARRIS'S BARK-LOUSE.

(*Aspidiotus Harrisii,* Walsh.)

This appears upon the trunks of small trees, and the branches of older ones, in the form of dirty-white scales,

Fig. 98.—HARRIS'S BARK-LOUSE.

which are usually irregularly egg-shaped; but, however variable in outline, it is always quite flat and causes the infested tree to wear the appearance of figure 98; while the minute eggs which are found under it in winter time are invariably blood-red or lake-red. This species has scarcely ever been known to increase sufficiently to do material damage, for the reason, doubtless, that there have, hitherto, always been natural enemies and parasites enough to keep it in due bounds.

THE OYSTER-SHELL BARK-LOUSE.

(*Mytilaspis pomicorticis*, Riley.)

The Oyster-shell Bark-louse, was formerly known as *Aspidiotus conchiformis*, but changed by Prof. Riley for good reasons to the name given above. It is one of the most pernicious and destructive insects with which the apple-grower in the Northern States has to contend. This species presents the appearance of figure 99, and may always be distinguished from the preceding, by having a very uniform mussel-shaped scale of an ash-gray color (the identical color of the bark), and by these scales, containing, in the winter time, not red, but pure white eggs.

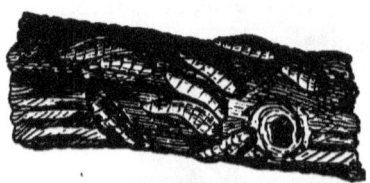

Fig. 99.—OYSTER-SHELL BARK-LOUSE.

There is scarcely an apple-orchard in Northern Illinois, in Iowa, or in Wisconsin, that has not suffered more or less from its attacks, and many an one has been slowly but surely bled to death by this tiny sap-sucker. It was introduced into the Eastern States about the beginning of the present century, from Europe, and had already reached as far west as Wisconsin in 1840, from whence it spread at a most alarming rate throughout the districts bordering on Lake Michigan. It occurs at the present time in Minnesota and Iowa, but whether or not it extends westward beyond the Missouri River, there are no data to show. Its extension southward is undoubtedly limited, for though so abundant in the northern half of

Illinois, observation has shown that it does not exist in the southern half of the same State.

As the female Bark-louse is only capable of motion for two to three days at the most, after which time she becomes as permanently fixed for the rest of her life, as is the tree on which she is fastened, it may puzzle some to divine how this insect spreads from tree to tree, and place to place. That it is transported to distant places, mainly on young trees, there can be no doubt, and there are various ways in which it can spread from tree to tree in the same orchard, though it can only thus spread during the few days of its active larval state.

Though essentially belonging to the Apple Tree, this Oyster-shell Bark-louse, is found upon the Currant, the Plum, and the Pear. I have seen the scales fully developed, and bearing healthy eggs on the fruit of the White Doyenne Pear, of the Transcendent Crab, and of Wild Plum (*Prunus Americana*); and, though on the hard bark of a tree, we cannot judge of the amount of sap they absorb, it is quite apparent on these soft fruits, for each scale causes a considerable depression from the general surface.

REMEDIES.—If an orchard is once attacked before the owner is aware of it, much could be done on the young trees by scraping the scales off in winter, but on large trees, where it is difficult to reach all the terminal twigs, this method becomes altogether impracticable, and it will avail but little to cleanse the trunk alone, as most of the scales containing living eggs will be found on the terminal branches. Alkaline washes, and all other washes, except those of an oily nature, such as petroleum and kerosene, are of no avail when applied to the scales, for the simple reason that they do not penetrate and reach the eggs which are so well protected by these scales; and it is very doubtful whether any solution can be used, that is sufficiently oily to penetrate the scales and kill the eggs without injury to the tree, especially while the sap of the tree is inactive. Hence the Bark-louse can only be successfully fought at the time the eggs are hatching, and the young lice are crawling over the limbs. The time of year in which this occurs, are the last days of May and the first days of June, but without close scrutiny they will not be observed, as they appear like very minute, white, moving specks. While the young larvæ are thus crawling over the tree, they are so tender that they can be readily destroyed by simply scrubbing the limbs with a stiff brush.

With regard to washes to be used with a syringe, the late Dr. Jno. Kennicott used one pound of Sal. Soda, to one gallon of water with good effect; Mr. E. G. Mygatt, of Richmond, McHenry County, Illinois, has experimented with this insect for over twenty years, with the following result: Brine (2 quarts salt to 8 of water), kills the lice, but also the foliage and fruit. Tobacco-water (strong decoction), neither injures the foliage nor affects the lice. Weak Lye, while it kills the lice, will also somewhat affect the leaves. Lime-water kills about half the lice, and affects the leaves a little. Finally, a decoction of Quassia, though well known to be effectual for the common Plant-lice, has no effect on these Coccids. In short, we have abundant proof that neither Tobacco-water, nor strong Alkaline washes, have any effect on the young lice, though a strong solution of soap will kill them, and my experience the past season, with Cresylic Acid soap in other directions, leads me to strongly recommend it for this purpose. It will sometimes be necessary to repeat the wash, as the lice do not all hatch out the same day, though the period of hatching seldom extends over three days.

From the foregoing it is obvious that Bark-lice can only be successfully fought during three or four days of the year; how absurd and ridiculous then, are all the patent nostrums and compounds, which are continuously offered to the public as perfect "Bark-lice extinguishers," and which never mention this important fact.

One case was reported to the "American Agriculturist" a few years ago, by the owner of a Pear Tree badly covered with this Bark-louse. Painters were at work painting the house, and in a fit of desperation he took a brush and painted the tree from the ground to the end of the smallest branch, expecting of course to kill it. Much to his surprise, the tree pushed its shoots as readily as ever, and was perfectly free from the insect. Another case was reported in the same journal of the efficacy of Crude Petroleum, used in the same manner on young Apple trees. These however may be regarded as desperate cases, and are only given as hints.

THE APPLE-TREE TENT-CATERPILLAR.

(Clisiocampa Americana, Harr.)

What orchardist in the older States of the Union is not familiar with the white web-nests of this caterpillar? As they glisten in the rays of the spring sun before the trees have put on their full summer dress, these nests, which are then small, speak volumes of the negligence and slovenliness of the owner of the orchard, and tell more truly than almost any thing else why it is that he fails and has bad luck with his apple crop. Wherever these nests abound one feels morally certain that the borers, the Codling-moth, and the many other enemies of the apple tree, have full play to do as they please, unmolested and unnoticed by him whom they are ruining.

The small, bright and glistening web, if unmolested, is soon enlarged until it spreads over whole branches, and the caterpillars which were the architects, in time become moths, and they lay their eggs for an increased supply of nests another year.

This insect is so well known throughout the country that it is only necessary to give here the most prominent and important points in its history, the more especially

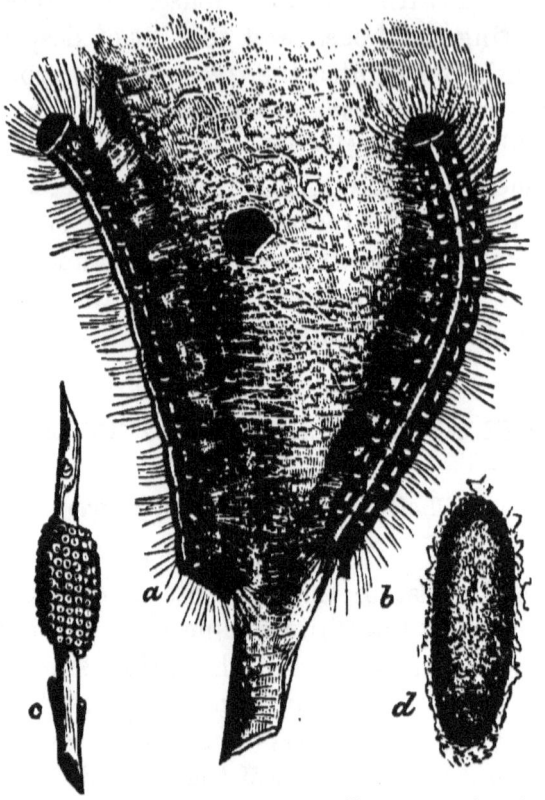

Fig. 100.—APPLE-TREE TENT-CATERPILLAR (*Clisiocampa Americana*, Harr.) *a*, Side View; *b*, Back View of Caterpillar; *c*, Eggs; *d*, Cocoon.

as the figures herewith given will alone enable the novice to recognize it the moment it appears in a young orchard.

The eggs (fig. 100, *c*), from which these caterpillars hatch are deposited mostly during the month of June, in oval rings, upon the smaller twigs, and this peculiar mode of deposition renders them conspicuous objects during the winter time, when by a little practice they can easily be distinguished from the buds, knots, or swellings of the naked twigs. Each cluster consists of from two to three hundred eggs, and is covered and protected from the weather by a coating of glutinous matter, and the same temperature which causes the apple-buds to swell and burst, quickens the vital energies of these larvæ and causes them to eat their way out of their eggs. Very often they hatch during a prematurely warm spell, and before there is any green leaf for them to feed upon, but they are so tough and hardy that they can fast for many days with impunity, and the glutinous substance on the outside of their eggs furnishes good sustenance and gives them strength at first. It is even asserted by Mr. H. C. Raymond, of Council Bluffs, Iowa, that the eggs often hatch in the fall, and that in these cases the larvæ withstand the severity of the winter with impunity.

Fig. 101.—APPLE-TREE TENT-CATERPILLAR, MOTH.

The young caterpillars commence spinning the moment they are born, and indeed they never move without extending their thread wherever they go. All the individuals hatched from the same batch of eggs work together in harmony, and each performs its share of building the common tent, under which they shelter when not feeding and during inclement weather. They usually feed twice each day, namely, once in the forenoon and once in the afternoon. After feeding for five or six weeks, during which time they change their skins four times, these caterpillars acquire their full growth, when they appear as at fig. 100 (*a* side view, *b* back view), the colors being black, white, blue and rufous or reddish. They then scatter in all directions in search of some cozy and sheltered nook, such as the crevice or angle of the fence, and having finally decided on the spot, each one spins an oblong-oval yellow cocoon (fig. 100, *d*), the silk composing which is intermixed with a yellow fluid or paste, which dries into a powder looking something like sulphur. A few individuals almost always remain and spin up in the tent, and these cocoons will be found intermixed with the black excrement long after the old tent is deserted.

Within this cocoon the caterpillar soon assumes the chrysalis state, and from it, at the end of about three weeks, the perfect insect issues as a dull yellowish-brown or a reddish-brown moth (fig. 101), characterized chiefly by the front wings being divided into three nearly equal parts by two transverse whitish or pale-yellowish lines, and by the middle space between these lines being paler than in the rest of the wing in the males, though it is more often of the same color, or even darker in the females. The species is, however, very variable.

The moths do not feed, and the sole aim of their lives seems to be the perpetuation of their kind; for as soon as they have paired and each female has carefully consigned her eggs to some twig, they die, and when the proper time comes around again the eggs will hatch, and the same cycle of changes take place each year.

This insect in all probability extends wherever the wild Black Cherry (*Prunus serotina*) is found, as it prefers this tree to all others; and this is probably the reason why the young so often hatch out before the apple buds burst, because, as is well known, the Cherry leaves out much earlier. Besides the Cherry and Apple, both wild and cultivated, the Apple-Tree Tent-caterpillar will feed upon Plum, Thorn, Rose, and perhaps on most plants belonging to the Rose family, though the Peach is not congenial to it, and it never attacks the Pear, upon which it is said that it will starve. It does well on Willow and Poplar, and even on White Oak, according to Fitch, who also found it on Witch Hazel (*Hamamelis*) and Beech.

REMEDIES.—No insect is more readily kept in subjection than this. Cut off and burn the egg-clusters during winter, and examine the trees carefully in the spring for the nests from such clusters as may have eluded the winter search. The eggs are best seen in a dull day in winter when they show distinctly against the sky. Though to kill the caterpillars numerous methods have been resorted to, such as burning, and swabbing with oil, soap suds, lye, etc., they are all unnecessary, for the nests should not be allowed to get large, and if taken when small are most easily and effectually destroyed by going over the orchard with the fruit-ladder, and by the use of gloved hands. As the caterpillars feed twice a day, once in the forenoon and once in the afternoon, and as they are almost always in their nests till after nine A. M., and late in the evening, the early and late hours of the day are the best in which to perform the operation. As a means of facilitating their destruction, it would be a good plan, as Dr. Fitch has suggested, to place a few Wild Cherry trees in the vicinity of the orchard, and as the moths will mostly be attracted to such trees to deposit their eggs, and as a hundred clusters on a single tree are destroyed more easily than if they were scattered over a hundred trees, these trees will repay the trouble wherever the Tent-caterpillar is known to be a grievous pest.

THE TENT-CATERPILLAR OF THE FOREST.

(*Clisiocampa sylvatica*, Harr.)

The egg-mass from which the Tent-caterpillar of the Forest hatches (fig. 102, *a*, showing it after the young larvæ have escaped) may at once be distinguished from that of the common Tent-caterpillar by its being of a uniform diameter, and docked off squarely at each end. It is usually composed of about four hundred eggs, (the number in five masses ranging from three hundred and eighty to four hundred and sixteen). Each of the eggs composing this mass is of a cream-white color, 0.04 inch long and 0.025 inch wide, narrow and rounded at the attached end or base, gradually enlarging towards the top, where it becomes slightly smaller (fig. 102, *d*), and abruptly terminates with a prominent circular rim on the outside, and a sunken spot in the center (*c*). These eggs are deposited in circles, the female moth stationing herself, for this purpose, in a transverse position across the twig. With abdomen curved she gradually moves as the depo-

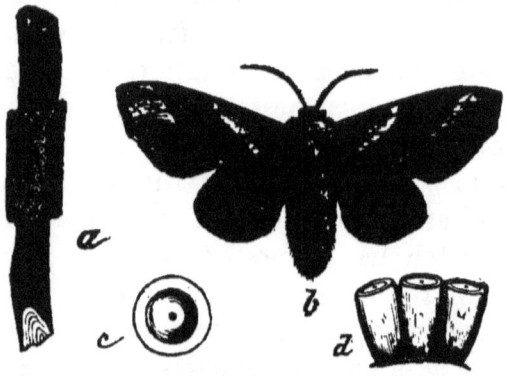

Fig. 102.—TENT-CATERPILLAR OF THE FOREST (*Clisiocampa sylvatica*).
a, Egg-mass; *b*, Moth; *c*, top of Egg; *d*, Eggs.

sition goes on, and when one circle is completed, she commences another—and not before. With each egg is secreted a brown varnish which firmly fastens it to the twig and to its neighbor, and which, upon becoming dry, forms a net-work of brown over the pale egg-shell. These eggs are so regularly laid and so closely glued to each other, and the sides are often so appressed, that the moth economizes space almost as effectually as does the Honey-bee in the formation of its hexagonal cells.

The eggs are deposited, in the latitude of St. Louis, during the latter part of June. The embryo develops during the hot summer weather and the yet unborn larva is fully formed by the time winter comes on. They hatch with the first warm weather, in spring—generally from the middle to the last of March—and though the buds of their food-plant may not have opened at the time, and though it may freeze severely afterwards, yet these little creatures are wonderfully hardy, and can fast for three whole weeks, if need be, and withstand any amount of inclement weather. The very moment these little larvæ are born, they commence spinning a web wherever they go. At this time they are black with pale hairs, and are always found either huddled together or travelling in file along the silken paths which they form when in search of food. In about two weeks from the time they commence feeding they go through their first moult, having first grown paler or of a light yellowish-brown, with the extremities rather darker than the middle of the body, with the little warts which give rise to the hairs quite distinct, and a conspicuous dark interrupted line each side of the back. After the first moult, they are characterized principally by two pale-yellowish subdorsal lines, which border what was before the dark line above described. After the second moult, which takes place in about a week from the first, the characteristic pale spots on the back appear, the upper pale line becomes yellow, the lower one white, and the space between them bluish: indeed, the characters of the mature larva are from this period apparent. Very soon they undergo a third moult, after which the colors all become more distinct and fresh, the head and anal plate have a soft bluish velvety appearance, and the hairs seem more dense. After undergoing a fourth moult without material change in appearance, they acquire their full growth in about six weeks from the time of first feeding. At this time they appear as at figure 103.

At this stage of its growth the Tent-caterpillar of the Forest may be seen wandering singly over different trees, along roads, on the tops of fences, etc., in search of a suitable place to form its cocoon. It usually contents itself with folding a leaf or drawing several together for this purpose, though it frequently spins up under fence boards and in other sheltered situations. The cocoon is much like that of the common Tent-caterpillar, being formed of a loose exterior covering of white silk with the hairs of the larva interwoven, and by a more compact oval inner pod that is made stiff by the meshes being filled with a thin yellowish paste from the mouth of the larva, which paste, when dried, gives the cocoon the appearance of being dusted with powdered sulphur. Three days after the cocoon is completed the caterpillar casts its skin for the last time and becomes a chrysalis of a red-

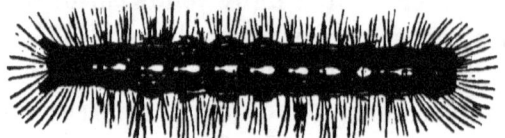

Fig. 103.—TENT-CATERPILLAR OF THE FOREST.

dish-brown color, slightly dusted with a pale powder, and densely clothed with short pale yellow hairs, which at the blunt and rounded extremity are somewhat larger and darker. In a couple of weeks more, or during the forepart of June, the moths commence to issue, and fly about at night. This moth (fig. 102, *b*, female), bears a considerable resemblance to that of the Common Tent-caterpillar (fig. 101), being of a brownish yellow or rusty brown, and having two oblique transverse lines across the front wings. It differs, however, in the color being paler or more yellowish, especially on the thorax; in the space between the oblique lines being usually darker instead of lighter than that on either side; but principally in the oblique lines themselves being dark instead of light, and in a transverse shade, often quite distinct, across the hind wings. As in *C. Americana*, the male is smaller than the female, with the wings shorter and cut off more squarely. Considerable variation may be found in a given number of moths, but principally in the space between the oblique lines on the front wings being either of the same shade as the rest of the wing, or in its being much darker; but as we have found these variations in different individuals of the same brood, bred either from Apple, Oak, Hickory, or Rose, they evidently have nothing to do with the food-plant. The scales on the wings are very loosely attached, and rub off so readily that good specimens of the moth are seldom captured at large

THE LARVA SPINS A WEB.

From the very moment it is born until after the fourth or last moult, this caterpillar spins a web and lives more or less in company; but from the fact that this web is always attached close to the branches and trunks of the trees infested, it is often overlooked, and several writers have falsely declared that it does not spin. At each successive moult all the individuals of a batch collect and huddle together upon a common web for two or three days, and during these periods—though more active than most caterpillars in this so-called sickness—they are quite sluggish. During the last or fourth moult they very frequently come low down on the trunk of the tree, and, unwittingly court destruction by collecting in masses within man's reach.

REMEDIES.—From their birth until after the third moult these worms will drop and suspend themselves in mid-air, if the branch upon which they are feeding be suddenly jarred. Therefore when they have been allowed to multiply in an orchard this habit will suggest various modes of destroying them. Again, as already stated, they can often be slaughtered *en masse* when collected on the trunks during the last moulting period. They will more generally be found on the leeward side of the tree if the wind has been blowing in the same direction for a few days. The cocoons may also be searched for, and many of the moths caught by attracting them towards the light. But the most effective artificial mode of preventing this insect's injuries is to search for and destroy the egg-masses in the winter time when the trees are leafless.

SUMMARY.

The Tent-caterpillar of the Forest differs from the common Orchard Tent-caterpillar principally in its egg-mass being docked off squarely instead of being rounded at each end; in its larva having a row of spots along the back instead of a continuous narrow line, and in its moth having the color between the oblique lines on the front wings as dark or else darker, instead of lighter than the rest of the wing. It feeds on a variety of both orchard and forest trees; makes a web which, from its being usually fastened close to the tree, is often overlooked; is very destructive, and is most easily fought in the egg state.

THE FALL WEB-WORM.

(Hyphantria textor, Harris.)

The appearance of webs, or "tents," upon fruit and other trees in late summer and early autumn, has caused many to suppose that there was a second brood of the Tent-caterpillar. These late webs belong to a very different insect, which lays her eggs in a cluster upon a leaf near the end of a twig, and the young caterpillars, like those of the true tent-makers, begin to spin as soon as hatched; and as they feed and spin in company, the web formed by their united efforts soon becomes conspicuous. The worms descend the branches, devouring the pulpy portions of the leaves upon them, and form a web as they go. When they have made their growth, the caterpillars descend to the ground, where, just beneath the surface, they enter the pupa state; the next summer they issue as pure white moths, to lay eggs for another brood. The worm, or caterpillar, is of a general pale-yellow color, with a broad dusky stripe along the back, and a yellow stripe along each side, and they have numerous whitish hairs. While the Fall Web-worm often attacks the Apple and other fruit trees, it does not confine itself to the orchard, but its webs may be seen in autumn upon various kinds of trees, as well as on shrubs. The only remedy is to destroy the web wherever it may be seen; and as the worms never leave the nest, this is quite sure to be effective.

THE APPLE-WORM—CODLING MOTH.

(Carpocapsa pomonella, Linn.)

This is one of the most important of the insects of the orchard, in view of the great loss it annually causes. While all those who eat apples have seen its work, a burrowing at the core of the fruit and an abundant deposit of excrement, very few, even among fruit growers, have seen the perfect insect, which is a small moth. Like most of our worm insect foes, it was originally a denizen of the Old World, having been introduced into this country only about the beginning of the present century. Twenty years ago it was unknown in Illinois; and it is only within the last eight or ten years that it has penetrated into Iowa.

The Apple-worm moth makes its appearance in North Illinois from the last of May to the forepart of June, and a little earlier or later according to the season and the latitude. Usually, at the time it appears, the young apples are already set, and beginning to be about as large as a hazel-nut. After coupling in the usual manner, the female moth then proceeds to deposit a single egg in the blossom end of the fruit, flying from fruit to fruit until her stock of eggs (amounting to probably two or three hundred) is exhausted. Not long after accomplishing this process she dies of old age and exhaustion. In a short time afterwards the egg, no matter where it is located, hatches out, and the young larva forthwith pro-

Fig. 104.—APPLE-WORM—CODLING MOTH (*Carpocapsa pomonella*, Linn.) Perfect Insect; Larva and its work; Pupa at the lower right-hand side.

ceeds to burrow into the flesh of the apple, feeding as it goes, but making its head-quarters in the core. In three or four weeks time it is full grown, and shortly before this, the infested apple generally falls to the ground. The larva then crawls out of the fruit through a large hole in the cheek, which it has bored several days beforehand for that express purpose, and usually makes for the trunk of the tree, up which it climbs, and spins around itself a silken cocoon of a dirty-white color, in any convenient crevice it can find, the crotch of the tree being a favorite spot. Here it transforms into the pupa state; and, towards the latter end of July or the forepart of August, bursts forth in the moth state. The different stages in the life of this insect are given in figure 104. The channel made by the young larva in reaching the core, and the cavity it makes in feeding there, are shown. At the upper right hand the full grown larva is given, and at the lower right-hand the pupa. At the left-hand side the perfect insect is shown, with its wings open and closed. The moth is distinguished from all other moths by a patch of coppery scales at the tip of its front wings.

The infested fruit does not always drop when the borer leaves it; seeks a place in which to undergo its changes, and in from ten to fifteen days a second brood of moths issues, and the fruit is re-stocked with larvæ. The second brood do not issue as moths until the next spring, many of the larvæ of the late brood do not leave the apples until they are harvested, and undergo their changes in the cellar.

REMEDIES.—The utility of pasturing swine in the orchard is generally admitted and did all the infested apples fall would be more than the partial remedy that it it now is. Acting upon the fact that many of the worms after leaving the fallen fruit seek a place of concealment upon the trunk in which to pupate, Codling-moth traps have been invented, and some have been patented. One of the most effective traps is a strip of carpet or other coarse woollen fabric, about five inches wide and long enough to go around the tree; this is fastened by a few tacks, which should not be driven home, as they need to be removed. These strips are examined every ten days and the insects killed. In large orchards the killing is expeditiously done by running the cloths between the rollers of a clothes wringer. Fruit cellars, and any empty boxes or barrels they may contain, should be examined before May for concealed pupæ.

THE APPLE-MAGGOT.

(*Trypeta pomonella*, Walsh.)

Besides the well-known Apple-worm, or Codling-moth, there is in some localities, especially in the older States, the Apple-maggot. It differs from the Codling-moth in many respects; the parent insect is not a moth, but one of the two-winged flies. It is not, like the other, an

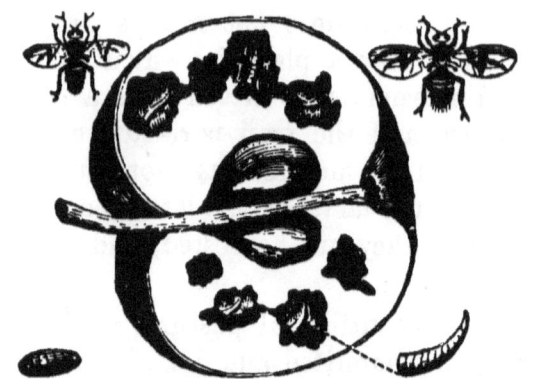

Fig. 105.—APPLE-MAGGOT (*Trypeta pomonella*, Walsh.)
Perfect Insect; Larva and its burrows; Pupa.

imported insect, but a native which has long inhabited our wild apples and the haws, or fruit of our thorns, and is found in cultivated fruit, here and there, all over the country. Figure 105 shows an infested apple, and the insect in its different stages, the perfect fly, with its transparent wings, being shown above, while the maggot and pupa are given below. The excavations in the apple show that the larvæ enter at no particular place, and do not, as in the case of the Codling-moth, seek the core. The destruction of the infested fruit by feeding it to pigs, or making it into cider, are among the obvious means of prevention.

THE APPLE CURCULIO.

(Anthonomus quadrigibbus, Say.)

Some have stated that the common Plum Curculio will also attack young Apples; however this may be, there is, in several of the Western States, and in Canada, a Curculio which has long infested the native Crab Apples, and has, in many cases, learned to prefer the cultivated to the wild fruit. A comparison of the engraving of this insect (fig. 106), with that of the Plum Curculio, given on a subsequent page, will at once show striking difference. In the first place, there is the greater length of snout, which is carried extended in front; then the marked widening of the body behind, serves also to distinguish it.

Fig. 106.—APPLE CURCULIO (*Anthonomus quadrigibbus*, Say.) *a*, Real size; *b*, Side view; *c*, Back view, both enlarged.

It has four conspicuous lumps on the wing-cases at the rear, from which it takes its specific name. It varies from one-twentieth, to one-twelfth of an inch in length. It is of a rusty-brown color, and the thorax, and often the forward third of the wing-covers ash-gray.

The insect deposits its egg in an opening made in the skin of the fruit; the larva when hatched goes to the core, and there feeds, producing much excrement, for nearly a month, and then assumes the pupa state within the fruit, which does not fall; in two or three weeks it appears as a perfect beetle. In Missouri and Southern Illinois, this insect often does much damage to the Apple crop, and probably it is abundant in other States, where its work has been attributed to other insects. In several cases it has been known to attack Pears as well as Apples.

REMEDIES.—It is not probable that much can be done with jarring down this insect, as advised for the Plum Curculio, as it is not like that easily alarmed. So far as known, it can only be attacked while within the fruit. Shaking or jarring the tree may be useful in bringing down the infested apples, which should be at once fed to swine, or otherwise destroyed.

THE CANKER-WORM.

(*Anisopteryx vernata*, Peck.)

The greatest injury done by Canker-worms is to Apple-trees, but it also attacks other fruit trees, and often injures shade trees, especially the Elm, which in some localities it completely defoliates. The male moth (fig. 107) has an expanse of wings of about an inch and a quarter; the wings are very thin and silky, the fore-wings ash-colored, with a small but distinct whitish spot on the front edge, near the tip; the hind wings are pale ash-colored. The female (fig. 108) is entirely wingless and of a general ash-gray color. Being without wings, she can only reach the branches of the tree to deposit her eggs, by crawling up the trunk, which she does very early in the spring; in mild weather even in February. The eggs are deposited in clusters of one hundred or more on the bark of the branches and twigs, and may often be found on the inside of the loose scales of the bark. When the leaves first begin to make their appearance, these eggs hatch into tiny Span-worms, scarcely visible to the naked eye, but they grow rapidly, and in three or four

Fig. 107.—MALE CANKER-WORM—MOTH (*Anisopteryx vernata*, Peck.)

weeks have attained their full size—about an inch in length, when they cease eating, and let themselves down by a silken thread and enter the ground, where they soon become chrysalids, in which state they remain all through the summer and fall, and usually until the following spring, when they emerge as moths. The fact that the female moth is wingless makes it a comparatively easy matter to keep these Canker-worms in check, for the parent moth is obliged to crawl up the trunk of the tree to deposit her eggs, and if she can be prevented from doing this, of course she must lay her eggs below the obstruction, where they can be easily destroyed.

Dr. William Le Baron gives the following remedies in his Second Illinois Report:

"1st. Prevent the passage of the moths up the trees. The most approved plan heretofore used is to put a canvas or other cloth band, six inches or more in width, around the trunk and besmear it with tar, or a mixture of tar and molasses, applied every other day. The method suggested in this Report is to put a band of rope or closely twisted hay around the trunk, and over this a tin band about four inches wide, so placed that the rope shall be at the middle of the tin, making a closed cavity below, and a free edge of tin above. The time to use these appliances is mostly in the month of March, but also at other times when the weather is sufficiently open to permit the insects to run.

Fig. 108.—FEMALE CANKER-WORM—MOTH.

"2nd. If the moths are prevented from ascending the tree, they will deposit their eggs below the obstruction, and for the most part near to it. These eggs can be destroyed by a single application of kerosene oil.

"3rd. If the moths are not prevented from ascending the tree, they will deposit their eggs mostly upon the underside of the scales of bark, on the upper part of the trunk and larger branches. Many of these can be destroyed by scraping off and burning the scales.

"4th. If all precautions have been neglected and the eggs have been permitted to hatch, then, as soon as the worms are large enough to be easily seen, jar them from the trees and sweep them away with a pole, as they hang by their threads, and burn or otherwise destroy them.

"5th. If the worms have matured and gone into the ground for winter quarters, plow the ground late in the fall, so as to expose the pupæ to frost, and to the action of natural enemies."

The rope and tin bands mentioned in the first paragraph are deserving of particular attention, as they have been found to be an almost perfect barrier to the ascent of the moths. The method of putting on these bands is very simple. Take a piece of inch rope—old worn out rope is as good as new—tack one end to the trunk, two feet or less from the ground, with a shingle nail, driven in so that the head shall not project beyond the level of the rope. Bring the rope around the tree, and let it lap by the beginning an inch or two, cut it off and fasten it in the same manner. Get the tinman to cut up some sheets of tin into strips four inches wide and fasten them together endwise, so that they shall be long enough to go around the trees over the rope band, having the rope at the middle. Let the ends of the tin lap a little, punch a hole through them and fasten them with a nail driven through the tin and rope into the tree. The result of this contrivance is, that the moths congregate below the obstruction, and sometimes pile up so as to go over on the tin. But when they reach the upper edge of the tin they go round and round until they become discouraged. A great deal of ingenuity has been displayed in the contrivance of barriers of various kinds for preventing the female Canker-worm moth from ascending the trees. A pin-maker in Connecticut made a barrier of several rows of pins thrust through a rubber band; this was to be put around the trunk with the points of the pins outward. Other devices consist of troughs of sheet lead to surround the trunk, with a channel in which some kind of oil may be placed. In all such cases the simplest methods are the best. In New Haven and other New England places, which pride themselves upon their fine elms, trees which the Canker-worm particularly infests, the chief reliance is upon bands of thick paper placed around the trunks; this has placed upon it a barrier of pine tar or of old printer's-ink. Whatever barrier is used, it requires frequent attention. All liquids like oil, or viscid materials like tar, etc., may be covered by blowing dust, leaves, etc., to form a bridge across them; indeed the insects themselves, being arrested, often form a bridge with their dead bodies for the passage of their successors, and during the season such barriers should be daily looked to and renewed if necessary.

It may be added that some orchardists, instead of using preventive measures, allow the insects to deposit their eggs on the trees, and then, when the caterpillars begin their work upon the foliage, destroy them by the use of Paris Green mixed with water, and thrown into the trees by means of a force-pump.

NOTE.—While the foregoing insects attack the Apple in preference to other fruit trees, they are occasionally, as mentioned under each, injurious to other trees. When we recollect that all our fruit trees belong to the same botanical family (the *Rosaceæ*), it will not be surprising to find an insect attacking several different trees indiscriminately. This large family is divided by botanists into several sub-families, one of which, the Almond Sub-family (*Amygdaleæ*), includes, what are popularly known as "Stone-fruits,"—Peach, Plum, Cherry, etc; another, the Pear Sub-family (*Pomeæ*), includes the Apple, Pear, Quince, etc., and it is not often that the insects which prey upon one sub-family attack the other. Still there are a few general feeders, which are injurious to nearly all fruit trees, and make it difficult to classify insects according to the trees upon which they feed. The insects which follow, while they also injure the Apple, do not confine themselves to it; some attack all fruit trees alike, while the Peach-borer and Plum Curculio restrict themselves to the stone-fruits.

THE RED-HUMPED CATERPILLAR.

(*Notodonta concinna*, Smith.)

Young Apple and Pear trees, and sometimes other fruit trees, are frequently defoliated, or have large branches completely stripped of their leaves in late summer or early autumn, by the Red-humped Caterpillar.

Fig. 109.—RED-HUMPED CATERPILLAR.
(*Notodonta concinna*, Smith.)

Fig. 110.—PUPA OF RED-HUMPED CATERPILLAR.

The eggs are usually deposited in July, in clusters on the underside of a leaf near the end of a branch, and the young caterpillars eat downward, making clean work of the foliage as they descend. The full-grown caterpillars (fig. 109), are an inch and a quarter long; the general color yellowish-brown, paler on the sides, and striped length-wise with slender black lines; the head is coral-red, and on the top of the fourth ring is a bunch or hump of the same brilliant color; there are several short black prickles along the top of the back. The caterpillar tapers towards the tail, and this end is always elevated when it is at rest. When full grown, all the caterpillars of the same brood descend to the ground at the same time, seek a hiding place under leaves, or just below the surface of the soil, where they form cocoons, and assume the chrysalis state (fig. 110), in which they remain until the following June, when the perfect insect issues as a small, neat-looking moth of a general light-brown color, the fore-wings are dark-brown along the inner margin, with a dark-brown spot near the middle. The wings expand from an inch, to an inch and three-eighths. If these caterpillars are noticed when first hatched, they will be found all near together, and may be readily destroyed.

THE TWIG GIRDLER.

(*Oncideres cingulatus*, Say.)

Fig. 111.—TWIG-GIRDLER (*Oncideres cingulatus*, Say.)

This beetle is known to girdle a great number of different trees, among which may be mentioned Apple, Pear, Peach, and Plum, Hickory, Elm, Persimmon, and American Linden. Both sexes of the beetle feed upon the bark of the Hickory, but only the females, so far as we are aware, girdle the twigs. After partly girdling a particular twig she lays a number of eggs in the upper portion that has been killed, each egg being usually inserted just beneath a bud. Figure 111 shows the insect and her work. The twig usually, though not always, breaks off by the force of the wind during winter, and the larvæ flourish upon the dead wood as it lies upon the ground, burrowing just beneath the bark, and when very numerous leaving little else than the outer bark. The beetles do this work in the fall of the year. The young larva hatches and works a short distance into the twig before winter sets in, and continues working through spring and summer, transforming to pupa only towards autumn. Some writers have stated that two years are required for its development. While this may be true farther north it is not true of the latitude of St. Louis. The Insect has been found destructive in Pennsylvania, Indiana, and other Western States. Wherever its prunings are found, they should be gathered and burned.

NEW YORK WEEVIL.

(*Ithycerus Noveboracensis*, Forster.)

This large snout-beetle kills the twigs by gnawing off the tender bark, in the early part of the season before the buds have put out, and later in the year it destroys the tender shoots which start out from old wood, by entirely devouring them. It attacks, by preference, the tender growth of the Apple, though it will also make free with that of the Peach, Plum, and Pear, and probably of other fruit as well as of forest trees.

This beetle belongs to the same family as does the Plum Curculio; it is distinguished from most of the other snout-beetles by the antennæ or horns being straight instead of elbowed or flail-shaped as they are in the common Plum Curculio, for instance. The specific name *Noveboracensis* which means "of New York" was given to this beetle ninety-eight years ago, by Forster, doubtless because he received his specimens from New York. But like many other insects which have been honored with the name of some Eastern State, it is far more common in the Mississippi Valley than it is in the

State of New York, it being scarcely known as an injurious insect in the East. The general color of the beetle is ash-gray, marked with black as in the cut (fig. 112, *c*), and with the scutel or small semi-circular space immediately behind the thorax, between the wings, of a yellowish color. Its larval habits were for a long time unknown, but it was recently ascertained that it breeds in the twigs and tender branches of the Bur Oak; we have good reason to believe that it also breeds in those of the Pignut Hickory. The female, in depositing, first makes a longitudinal excavation with her jaws (fig. 112, *a*), eating upwards under the bark towards the end of the branch, and afterwards turns round to thrust her egg into the excavation. The larva (fig. 112, *b*), hatching from the egg is of the usual pale-yellow color with a tawny head. We have watched the whole operation of depositing, and, returning to the punctured twig a few days after the operation was performed, have cut out the young larva; but we do not yet know how long a time the larva needs to come to its growth, nor whether it undergoes its transformations within the branch, or leaves it for this purpose, to enter the ground; though the former hypothesis is the most likely.

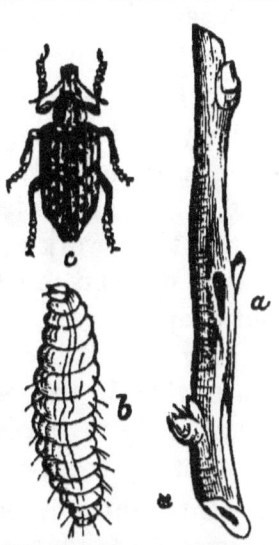

Fig. 112.—NEW YORK WEEVIL (*Ithycerus Noveboracensis*, Forster.)
a, Puncture; *b*, Larva; *c*, Beetle.

The same methods of catching this beetle may be employed as with the Plum Curculio.

CLIMBING CUT-WORMS.

Orchardists in spring frequently find the hearts of their fruit buds—on young trees especially—entirely eaten out and destroyed, and this circumstance is attributed to various causes, winged insects, beetles, slugs for instance; to birds or even to late frosts, when probably the entire mischief is caused by Cut-worms.

When climbing, Cut-worms will crawl up a tree eight or ten feet high, and seem to like equally well the leaves of the Pear, Apple, and Grape.

They work during the night, always descending to the earth again at early dawn, and hiding just under its surface, which accounts for their never having been noticed in this their work of destruction in former years. They seldom descend the tree as they ascend it, by crawling, but drop from the bud or leaf on which they have been feeding; and it is quite interesting to watch one at early morn when it has become full fed and the tender skin seems ready to burst from repletion, and see it prepare by a certain twist of the body for the fall.

"On light soil they often destroy low-branched fruit trees of all kinds, except the Peach, feeding on the fruit buds first, the wood buds as a second choice, tender grape buds and shoots (to which they are also partial), not excepted; the miller always prefers to lay her eggs near the hill or mound over the roots of the trees in the orchard; and if, as is many times the case, the trees have a spring dressing of lime or ashes with the view of preventing the May-beetle's operations, this will be selected with unerring instinct by the miller, thus giving her larvæ a fine warm bed to cover themselves up in during the day from the observations of their enemies. They will leave potatoes, peas, and all other green things for the Apple and Pear. The long, naked young trees of the orchard are almost exempt from their voracious attacks, but I have found them about midnight, of a damp and dark night well up in the limbs of these. The habit of the Dwarf Apple and Pear tree, however, just suits their natures, and much of the complaint of those people who cannot make these trees thrive on a sandy soil, has its foundation here, though apparently utterly unknown to the orchardist. There is no known remedy; salt has no properties repulsive to them, they burrow in it equally as quick as in lime or ashes. Tobacco, soap and other diluted washes do not even provoke them; but a tin tube six inches in length, opened on the side and closed around the base of the tree, fitting close and entering at the lower end an inch into the ground, is what the lawyers would term an effectual estoppel to further proceedings.

"If the dwarf tree branches so low from the ground as not to leave six inches clear of trunk between the limbs and ground, the limbs must be sacrificed to save the tree—as in two nights four or five of these pests will fully and effectually strip a four or five-year-old dwarf tree of every fruit and wood bud, and often when the tree is green, utterly denude it of its foliage. I look upon this Cut-worm as an enemy to the orchard more fatal than the Canker-worm, when left to themselves, but fortunately for mankind more surely headed off." J. W. Cochran, Calumet, Illinois.

The Climbing Cut-worm seems to prefer the Apple, Pear, and Grape-vine, though it also attacks the Blackberry, Raspberry, Currant, and even Rose-bushes and ornamental trees.

The subject is all important to the orchardist, and to those especially who have young and newly-planted trees on a light soil; for there are many who have had their trees injured by the buds being devoured in this manner, who never dreamed of preventing such an occurrence, for the reason that the mischief was attributed to birds.

Thus our quail, purple-finch, and many other birds, have too often unjustly received the execrations of the fruit culturist, which that evil genius, the Cut-worm, alone deserved. To understand an enemy's foible is to have conquered, and when we learn the source of an evil it need exist no longer. The range of these Climbing Cut-worms seems to be wide, for we have undoubted evidence of their attacking the Grape-vine in California, and I have found two species in Missouri, which have the same habit. Climbing Cut-worms frequently have the same habit of severing plants, as those which have never been known to climb, and I very much incline to believe that this habit is only acquired in the spring time, and most Cut-worms will mount trees if they are forced to do so, by the absence of herbaceous plants.

The Climbing Cut-worm (*Agrotis scandens*, Riley), has a similar general appearance to those which do not climb (see fig. 50, page 80). Its general color is a very light yellowish-gray, variegated with dirty bluish-green, and when filled with food it wears a much greener appearance than otherwise. In depth of shading it is variable, however, and the young worm is of a more uniform dirty whitish-yellow, with the lines along the body less distinct, but the shiny spots more so than in the full grown ones. Mr. Cochran informs us that on the Apple tree, when this worm has fed out its bud, the work is effectually done, that no adventitious or accessory bud ever starts again from the same place; the worm, as it were, boring into the very heart of the wood and effectually destroying the ability of the tree to re-act, at such a point, in the formation of a new bud, and that consequently a tree that is once stripped generally dies, and that this occurs more frequently on small or dwarf trees, where the buds are few, and three or four worms in a single night can eat out every one.

THE BAG-WORM, BASKET-WORM, or DROP-WORM.

(Thyridopteryx ephemeræformis, Haw).

The Bag-worm may be regarded as a Southern rather than a Northern insect, though it is found as far North as the northern part of New Jersey.

It is known to occur on Long Island, N. Y., in New Jersey, Massachusetts, New York, Pennsylvania, Ohio, Maryland, District of Columbia, the Carolinas, Georgia, Alabama, Kentucky, South Illinois, and South Missouri. Like the Canker-worm, the Tussock-moth, and all other insects in which the perfect female is wingless, the Bag-worm is extremely local in character, often abounding in a particular neighborhood, and being totally unknown a few miles away.

The clothing made by different insects, for protection either against the inclemency of the weather or against their enemies, is even more varied in cut and make-up, than are the divers costumes of the different peoples, civilized and barbarous, which inhabit our globe. Some insects live in the interior of leaves, using the upper and under cuticles as protection; some make their coats out of leaves themselves; some make cases of a sort of gummy cement, while others use cases of spun silk; but by far the greater number of those which protect themselves at all, employ silken cases which they cover and disguise with some other material. Thus lichens, grass, rushes, stones, shells, sand, wool, cotton, hair, wax, and the bark, twigs and leaves of trees, are all used for this purpose, while a few worms actually use their own excrement arranged on the outside of their cases with mathematical precision; unlike us mortals, however, these insects do not change the fashion of their dress with every change of season, but follow strictly the pattern used by their ancestors, who cut, spun, and wove, ages before our primordial mother sewed fig-leaves together. The follicle of our Bag-worm is covered by the leaves and stems of those trees or shrubs on which it subsists; and when evergreen leaves are used, they are often very regularly and prettily arranged after the fashion of thatching.

Throughout the winter, the weather-beaten bags of this insect may be seen hanging from almost every kind of tree; upon plucking them at that season many of them will be found empty, but the greater proportion of them will, on being cut open, be found partly full of soft yellow eggs. Those which do not contain eggs, are the male bags, and his empty chrysalis skin is generally found protruding from the lower end. From the middle to the end of May, in the latitude of St. Louis, these eggs hatch into little active brown worms, which, from the first moment of their lives, commence to form for themselves coverings. They crawl on to a tender leaf, and attached by the anterior legs, with their tails hoisted in the air, they each spin around themselves a ring of silk, to which they soon fasten bits of leaf. They continue adding to the lower edge of the ring, pushing it up as it increases in depth, until it reaches the tail, and forms a sort of cone, as represented in fig. 113, *g*. As the worms grow, they continue to increase their bags from the bottom, until the latter become so large and heavy that the worms allow them to hang, instead of holding them upright, as they did when they were young. By the end of July, the worms acquire their full growth, when they present the appearance of figure 113, *f*. At this stage, on being pulled out of its bag, or follicle, the worm appears as at fig. 113, *a*, that portion of the body which is always covered by the bag, being soft, and of a dull, smoky-brown, inclining to reddish at the sides; while the three anterior, or thoracic segments, which are exposed when the insect is feeding or marching, are horny, and mottled with black and white. The prolegs on the hidden part of the body are but poorly developed, and consist of but slight wart-like projections; they are furnished, however, with numerous small hooks, which answer an admirable purpose, in enabling the bearer to cling to his home-spun coat, which shelters him from the weather, and defends him from his enemies, and which is even more essential to his existence than are the clothes we wear to ours. The worms do not arrive at their full-grown condition without passing through critical periods. At four different times during their growth they close up the mouth

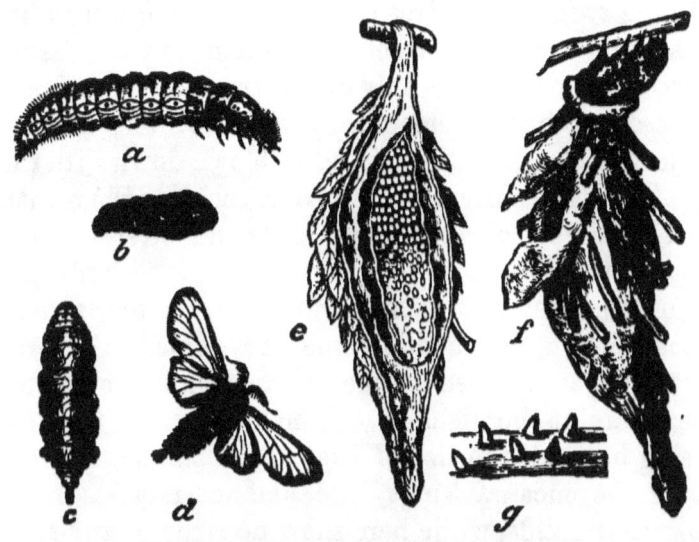

Fig. 113.—BAG, BASKET, OR DROP-WORM.
(*Thyridopteryx ephemeræformis*, Haw.)

a, Larva; *b*, Chrysalis; *c*, Female; *d*, Male; *e*, Female bag opened; *f*, The Worm and its Bag; *g*, The Young.

of their bags, and retire for two days to cast their skins or moult, as is the nature of their kind, and they push their old skins through a passage which is always left open at the extremity of the bag, and which also allows them to throw out their excrement.

During their growth they are very slow travellers, and seldom leave the tree on which they were born; but when full grown, they become quite restless; and it is at this time that they wander in the day time, dropping on persons by their silken threads, and crossing the sidewalks of our cities in all directions. It is from this habit of dropping upon persons, that they have been called "Drop-worms." A wise instinct urges them to thus wander from place to place, for, did they remain on the tree, they would soon multiply beyond the power of that tree to sustain them, and would in consequence become extinct. When they have lost their migratory desires, they fasten their bag very securely by a strong band of silk to the twigs of the tree on which they happen to be. Here again a strange instinct leads them to thus fasten their cocoons to the twigs only of the tree they inhabit, so that these cocoons will remain through the winter; and not to the leaf stalk, where they would be blown down with the leaf. After thus fastening their bags, they line them with a good thickness of soft white silk, and after turning around in the bag so as to have the head towards the lower orifice, they rest awhile from their labors, and at last cast their skins, and become chrysalids. Hitherto the worms had all been alike in appearance, but now the sexes are distinguishable, the male chrysalis (fig. 113, b), being but half the size of that of the female, and exhibiting the encased wings, legs, and antennæ, as in all ordinary chrysalids, while hers show no signs of any such members (see inside of bag at e). Three weeks afterwards, a still greater change takes place, the sexes differentiating still more. The male chrysalis works himself down to the end of his bag, and, hanging half-way out, the skin bursts, and the moth (fig. 113, d), with a black body and glassy wings, escapes, and, when his wings are dry, soars through the air to seek his mate, who is not blessed with wings, but is an abortive affair, with the head and general appearance of the larva, but still more degraded, since she has not even the legs which it possessed; she is, in fact, a naked, yellowish bag of eggs, with a ring of soft, light-brown, silky hair near the tail. (See fig. 113, c).

The female never withdraws herself entirely from the pupa shell, but holds on to it by her terminal segments, being evidently assisted by the ring of woolly hair already referred to. Thus, with the pupa shell extended to its utmost capacity, and the additional length of her whole body, she is enabled to reach to the lower orifice of the follicle, where she pertinently awaits the male, and after meeting him, works herself back into the pupa shell. Here she deposits her eggs in the upper part, intermingling them, and crowding the lower part of the puparium with the peculiar fawn-colored down already referred to. After having thus cosily secured her eggs against the winter's blasts, she works herself out and drops exhausted to the ground.

This insect is a general feeder, for it occurs alike on evergreen and deciduous trees. We have found it on the Apple, Plum, Cherry, Quince, Pear, Red and White Elms, the common Black and Honey Locusts, Lombardy Poplar, Catalpa, Norway Spruce, Arbor-vitæ, Osage Orange, Soft and Silver Maples, Sycamore, Linden, and above all, on the Red Cedar, while Mr. Glover has also found it on the Cotton plant in Georgia. We have even seen the bags attached to Raspberry canes.

This insect is also exceedingly hardy and vigorous, and the young worms will at first make their bags of almost any substance upon which they happen to rest, when newly hatched. They will construct them of leather, paper, straw, cork, wood, or of any other material which is sufficiently soft to allow of their gnawing it, and it is quite amusing to watch their operations.

REMEDIES.—How often does the simple knowledge of an insect's habits and transformations, give the clue to its easy destruction! From the foregoing account of the Bag-worm, it becomes obvious, that by plucking and burning the cases in winter, the trees can be easily rid of them. If this is done whenever the first few bags are observed, the task of plucking is light; but where it is not so done, the worms will continue to increase, and partly defoliating the tree each year, slowly, but surely, sap its life.

THE SLUG OF THE PEAR AND CHERRY TREE.

(Selandria cerasi, Peck.)

In New England, in June and July, there appears upon the leaves of the Pear and Cherry dark-green slimy creatures, so unlike caterpillars in general that they have received the popular name of slug. When grown, they are nearly half an inch ($9/_{20}$) long; being largest before and tapering behind, they have something of a tadpole appearance; the head is concealed under the fore-part of the body, and they usually have the tail somewhat turned up when at rest. Their color is a dark-blackish or bottle-green, and they exude from their skins a slimy matter which forms a shining trail wherever they move. They eat away the pulpy substance of the leaf, completely skeletonizing it; and as there are sometimes as many as twenty or more on a single leaf, they may do much injury by defoliating the tree, and causing leaves to push out from the buds prepared for next season. When present in large numbers, they give off an unpleasant odor, which may be noticed at some distance from the trees. It takes these slugs about twenty-six days to complete their growth, and after their final moult, they no longer have their slug-like appearance, but, as clean yellow caterpillars, leave the trees, and entering the ground for a few inches, form an oval earthen cocoon, in which they become chrysalids, and at the end of sixteen days come out in their perfect state, that of a fly of the order *Hy-*

menoptera. The female fly is slightly over one-fifth of an inch long, the male somewhat smaller, of a glossy-black, the first two pairs of legs being yellowish, with blackish thighs. The transparent wings are iridescent, the front pair having a smoky tinge across the middle. They lay their eggs in little incisions made in the skin of the leaf. This insect is often very injurious in the older States, and is extending westward, and is frequent in Canada.

REMEDIES.—Dry air-slaked lime, if sprinkled from a perforated tin vessel, or from a bag of some open fabric, attached to a pole, has been found very effective. It has been stated that the action of the lime is merely mechanical, and that fine dust, such as road-dust, will answer as well. The action of the dust may make the slug uncomfortable for awhile, but on shedding its skin it soon gets rid of it, while the lime soon kills the slug. Tobacco-water, Lime-water, and White Hellebore, used as directed under "Currant-worm," have been found of service.

THE PEACH-BORER.

(*Ægeria exitiosa*, Say.)

This borer is quite common, and the greatest insect enemy with which the Peach grower has to contend.

From the Round-headed Apple-Tree Borer, to which it bears some resemblance, both in its mode of work, and general appearance, it is at once distinguished by having six scaly, and ten fleshy legs. It works also more generally under the surface of the ground, and goes through its transformations within a year, though worms of two or three sizes may be found at almost any season. When full grown, the worm spins for itself a follicle of silk, mixed with gum and excrement, from which in due time issues a moth. The figures show, 114, the male, and 115, the female. As will be seen from these engravings, the two sexes differ very materially from each other, the general color in both being glossy steel-blue.

This Borer also attacks the Plum Tree, though sin-

Fig. 114.—PEACH-BORER—MALE.   Fig. 115.—PEACH-BORER—FEMALE.

gularly enough, it causes no exudation of gum in this, as it does in the Peach Tree.

REMEDIES.—As the borer often attacks the young trees in the nursery, all trees before planting should be carefully examined near the root, and if any are present, they may be readily cut out. In large peach orchards, "worming" is a part of the labor of cultivation. After the harvest, hands are employed to examine every tree for borers, and the more careful cultivators examine the trees in the spring also. The eggs are deposited from the middle of June, occasionally until October, at the surface of the ground. The grubs so soon as hatched, bore their way through the bark, and enter the sapwood. An exudation of gum at the base of the tree, is a sure sign of the presence of the borer. The earth is scraped away from the base of the tree, and a strong knife is used to cut away the dead and diseased bark and wood, and expose the hole; then a flexible probe, one of whale-bone is preferred, is thrust in to crush the borer. Sometimes as many as five or six are found in one tree, but all must be killed. After the operation the surface soil is drawn up to the tree to cover the wound. Boiling water applied to the base of the tree has been found useful. The borer may be prevented from laying her eggs, by surrounding the base of the tree with paper, which should extend for two inches below, and at least six inches above the surface of the ground, securing the upper portion by means of string or wire. Cloth and other preventives may be used in the same manner.

THE PLUM CURCULIO.

(*Conotrachelus nenuphar*, Herbst.)

The Plum Curculio, commonly known all over the country as THE Curculio, is a small, roughened, warty, brownish beetle, belonging to a very extensive family known as Snout-beetles (*Curculionidæ*). It measures about one-fifth of an inch in length, exclusive of the snout, and may be distinguished from all other North American Snout-beetles by having an elongate, knife-edged hump, resembling a piece of black sealing-wax, on the middle of each wing-case, behind which humps there is a broad clay-yellow band, with more or less white in its middle.

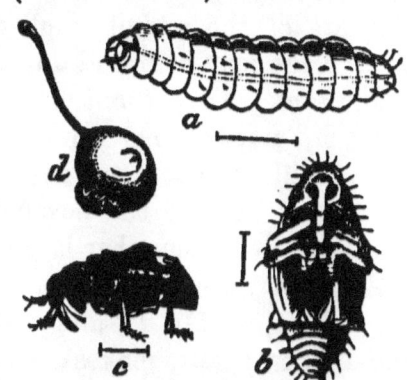

Fig. 116—PLUM CURCULIO.
(*Conotrachelus nenuphar*, Herbst.)
a, Larva; b, Pupa; c, Beetle; d, Beetle at work.

The engraving, figure 116, *c*, shows the magnified beetle, and at figure 117 it is represented at work, still more enlarged.

This is the perfect or imago form of the Curculio; and it is in this hard, shelly, beetle state, that the female passes the winter, sheltering under the shingles of houses, under the old bark of both forest and fruit trees, under logs and in rubbish of all kinds. As spring approaches, it awakens from its lethargy, and, if it has slept in the forest, instinctively searches for the nearest orchard. In Central Illinois and in Central Missouri the beetles may be found in the trees during the last half of April, but in the extreme southern part of Illinois they appear about two weeks earlier, while in the extreme northern part of the same State they are fully two weeks later. Thus, in the single State of Illinois, there is a difference of about one month in the time of the Curculio's first appearance on fruit trees; and the time will vary with the forwardness or lateness of the season.

As we shall see from the sequel, it is very important that we know just when first to expect "Mrs. Turk," and I therefore lay it down as a rule, applicable to any latitude, that she first commences to puncture peaches when they are of the size of small marbles or of hazel-nuts, though she may be found on the trees as soon as they are in blossom. To prevent confusion I will use the word "peach," not that her work is confined to this fruit, for, as we shall presently see, she is not so particular in her tastes, but because the peach is more extensively grown than are any of the other large kinds of stone-fruit.

Alighting, then, on a small peach, she takes a strong hold of it (fig. 116 d), and with the minute jaws at the end of her snout, makes a small cut just through the skin of the fruit. She then runs the snout slantingly under the skin, to the depth of one-sixteenth of an inch, and moves it back and forth until the cavity is large enough to receive the egg it is to retain. Then she turns around and drops an egg into the mouth of the cavity, and after this is accomplished, she resumes her first position, and by means of her snout pushes the egg to the end of the passage, and afterwards deliberately cuts the crescent in front of the hole, so as to undermine the egg and leave it in a sort of flap. The whole operation requires about five minutes, and her object in cutting the crescent is evidently to deaden the flap, so as to prevent the growing fruit from crushing the egg.

Now that she has completed this task, and has gone off to perform a similar operation on some other fruit, let us from day to day watch the egg which we have just seen deposited, and learn in what manner it develops into a Curculio like the parent which produced it—remembering that the life and habits of this one individual are illustrative of those of every other Plum Curculio.

We shall find that the egg is oval and of a pearly-white color. Should the weather be warm and genial, this egg will hatch in from four to five days, but if cold and unpleasant the hatching will not take place for a week or even longer. Eventually, however, there hatches from the egg a soft, tiny, footless grub with a horny head, and this grub immediately commences to feed upon the green flesh of the fruit, boring a tortuous path as it proceeds. It riots in the fruit—working by preference around the stone—for from three to five weeks, the period varying, according to various controlling influences.

Fig. 117.—THE PLUM CURCULIO.
The insect and its work, greatly enlarged.

The fruit containing this grub does not, in the majority of instances, mature, but falls prematurely to the ground, generally before the grub is quite full grown. I have known fruit to lie on the ground for upwards of two weeks before the grub left, and have found as many as five grubs in a single peach which had been on the ground for several days. When the grub has once become full grown, however, it forsakes the fruit which it has ruined, and burrows from four to six inches in the ground. At this time it is of a glassy yellowish-white color, though it usually partakes of the color of the fruit-flesh on which it was feeding. It is about two-fifths of an inch long, with the head light brown; there is a lighter line running along each side of its body, with a row of minute black bristles below, and a less distinct one above it, while the stomach is rust-red, or blackish. The full grown larva presents the appearance of figure 116, *a*.

In the ground, by turning round and round, it compresses the earth on all sides until it has formed a smooth oval cavity. Within this cavity, in the course of a few days, it assumes the pupa form, figure 116, *b*.

After remaining in the ground in this state for just about three weeks, it becomes a beetle, which, though soft and uniformly reddish at first, soon assumes its natural colors; and, when its several parts are sufficiently hardened, works through the soil to the light of day.

The Curculio when alarmed, like very many other insects, and especially such as belong to the same great Order of Beetles (*Coleoptera*), folds up its legs close to the body, turns under its snout into a groove which receives it, and drops to the ground. In doing this it feigns death, so as to escape from threatened danger, and does in reality very greatly resemble a dried fruit bud. It attacks, either for purposes of propagation or for food, the Nectarine, Plum, Apricot, Peach, Cherry, Apple, Pear and Quince, preferring them in the order of their naming.

It is always most numerous in the early part of the season on the outside of the orchards that are surrounded with timber. It is also more numerous in timbered regions than on the prairie.

It can fly and does fly, especially during the heat of the day; so cotton bandages around the trunk, and all like contrivances, are worse than useless.

It prefers smooth-skinned to rough-skinned stone-fruit.

The Miner Plum, otherwise known as the Hinckley Plum, and other varieties of that wild species known as the Chickasaw Plum (*Prunus Chicasa*), are less liable to its attacks than other kinds.

Both the male and female puncture the fruit for food, by gouging hemispherical holes; but the female alone makes the crescent-mark above described.

Scarcely any eggs are deposited after the stone of the fruit has become hard.

The cherry when infested remains on the tree, and the preventive measures that may be applied to other fruits will consequently not hold good with this.

The larva cannot well undergo its transformations in earth which is dry or baked, and severe drouths are consequently prejudicial to its increase.

It often matures in apples and pears, especially in early varieties, but in the great majority of instances the egg either fails to hatch or the young larva perishes in a few days after hatching.

ARTIFICIAL REMEDIES.—The remedies are few. They consist of prevention, by destroying the fallen fruit which contains the grub, and by jarring down and catching and killing the beetles. There are a variety of means which can be employed for destroying the grubs which fall with the fruit before they enter the ground. It can be done either by hand or by stock. Hogs and poultry are of undoubted use for this purpose. Of course, the first year they are used they do not in the least decrease the number of beetles, but wherever they can be used, a most beneficial effect will be noticed the second year, and every year afterwards. All attempts to repel the Curculio by hanging corn-cobs soaked in kerosene in the tree, or by throwing offensive mixtures upon the foliage have proved useless. The most effective method thus far discovered, is to jar down the insects and catch them on sheets. The tree should have a sudden jarring, not a mere shaking. For this purpose it is a good plan to saw off a small limb, leaving a stump a foot or less long, upon which to strike with a heavy mallet, this avoids bruising the bark of the tree. To catch the insects, two pieces of sheeting, each two yards long and a yard wide, may be stiffened by means of small rods or sticks, one at each long side and one in the middle; make the end of these sticks sharp, and cut a notch at a short distance from the end; the points of the sticks may be pushed into the cloth, and the notches will prevent that from slipping. A person can readily carry these from tree to tree, and placing them on the ground, one each side of the trunk, the tree is then to be jarred by a stroke of the mallet. The fallen insects may be crushed between the fingers, or be placed in a vessel of water, upon which there floats a small quantity of kerosene.

THE PERIODICAL OR SEVENTEEN-YEAR CICADA.

(*Cicada septendecim*, Linn.)

This insect is popularly known as the "Seventeen-year Locust," and by many confounded with the true Locust, and said to devour every green thing, whereas it is entirely unlike the "Rocky Mountain Locust," or "Colorado Grasshopper;" it belongs to another family, and has no jaws with which to devour anything. It simply has a beak to suck the juices of plants. It probably does its greatest injury in its under-ground life, sucking the juices of the roots of trees and plants, where it lives seventeen years. It however does much injury to fruit and other trees; the excavations made by the female form bad, ragged wounds, and sometimes so weaken the small branches that they are broken off. The following is mainly condensed from the description by Harris in his work on "Insects Injurious to Vegetation."

The Seventeen-year Cicada in the winged state (fig. 118, *c*), is of a black color, with transparent wings and

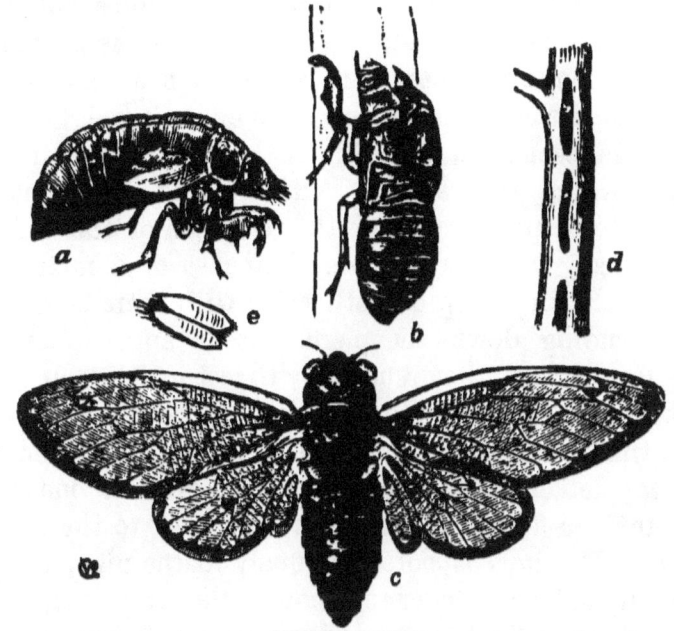

Fig. 118—PERIODICAL CICADA (*Cicada septendecim*, Linn.)
a, Pupa; *b*, Pupa Shell; *c*, Perfect Insect; *d*, Twig with Punctures, natural size; *e*, Eggs, enlarged.

wing-covers, the thick anterior edge and larger veins of which are orange-red, and near the tips of the latter there is a dusky zigzag line in the form of the letter W, supposed by the superstitious to indicate war; the eyes when living are also red; the rings of the body are edged with dull-orange; and the legs are of the same color. The wings expand from two and one-half to three and one-quarter inches.

In its many years of underground life this insect does more or less damage by feeding upon the roots of plants, but its manifest injury is only when, in the perfect state, the female deposits her eggs in the twigs of fruit trees, and at the times of periodical abundance the injury it causes is often serious, and it is properly classed among those injurious to fruit trees.

In those parts of the country which are subject to the visitation of this Cicada, it may be seen in forests of Oak through the month of June. And such immense numbers are sometimes congregated as to bend and even break down the limbs of the trees by their weight, and the woods resound with the din of their discordant drums from morning to evening. After pairing, the females proceed to prepare a nest for the reception of their eggs. They select, for this purpose, branches of a moderate size, which they clasp on both sides with their legs, and then, bending down the piercer at an angle of about forty-five degrees, they repeatedly thrust it obliquely into the bark and wood in the direction of the fibres, at the same time putting in motion the lateral saws, and in this way detach little splinters of the wood at one end, so as to form a kind of fibrous lid or cover to the perforation. The hole is bored obliquely to the pith, and is gradually enlarged by a repetition of the same operation, until a longitudinal fissure is formed of sufficient extent to receive from ten to twenty eggs. The side-pieces of the piercer serve as a groove to convey the eggs into the nest, where they are deposited in pairs, side by side, but separated from each other by a portion of woody fibre, and they are implanted in the limb somewhat obliquely, so that one end points upwards. When two eggs have been thus placed, the insect withdraws the piercer for a moment, and then inserts it again and drops two more eggs in a line with the first, and repeats the operation until she has filled the fissure from one end to the other, upon which she removes to a little distance, and begins to make another nest to contain two more rows of eggs. She is about fifteen minutes in preparing a single nest and filling it with eggs; but it is not unusual for her to make fifteen or twenty fissures in the same limb; and one observer counted fifty nests extending along in a line, each containing fifteen or twenty eggs in two rows, and all of

Fig. 119.—A PUNCTURED TWIG.

them apparently the work of one insect. After one limb is thus stocked, the Cicada goes to another, and passes from limb to limb and from tree to tree, until her store, which consists of four hundred or five hundred eggs, is exhausted. At length she becomes so weak by her incessant labors to provide for a succession of her kind, as to falter and fall in attempting to fly, and soon dies. Figure 118, d, shows a twig in which the eggs have been laid, and another is given in figure 119.

Although the Cicadas abound most upon the Oak, they resort occasionally to other forest trees, and even to shrubs, when impelled by the necessity for depositing

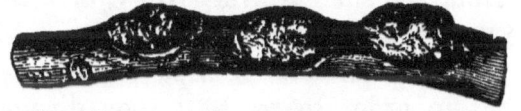

Fig. 120.—TWIG WITH HEALED PUNCTURES.

their eggs, and they very often commit them to fruit-trees, when the latter are in their vicinity. The punctured limbs languish and die soon after the eggs which are placed in them are hatched; they are broken by the winds or by their own weight, and either remain hanging by the bark alone, or fall with their withered foliage to the ground. In this way orchards have suffered severely in consequence of the injurious punctures of these insects. Sometimes, however, the twigs of the Apple and other fruit trees recover from these attacks and new wood forms over the wounds as shown in figure 120.

The eggs (fig. 118, *e*) are one-twelfth of an inch long, and one-tenth of an inch through the middle, but taper at each end to an obtuse point, and are of a pearl-white color. The shell is so thin and delicate that the form of the included insect can be seen before the egg is hatched. The young insect when it bursts the shell is one-sixteenth of an inch long, and is of a yellowish-white color, except the eyes and the claws of the fore legs, which are reddish; and it is covered with little hairs. In form it is some-

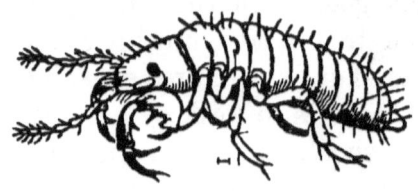

Fig. 121.—LARVA, MUCH ENLARGED.

what grub-like, being longer in proportion than the parent insect, and is furnished with six legs, the first pair of which are very large, shaped almost like lobster claws, and armed with strong spines beneath, On the shoulders are little prominences in the place of wings; and under the breast is a long beak for suction. Fig. 121 shows the larva enlarged. These little creatures when liberated from the shell are very lively, and their movements are nearly as quick as those of ants. After a few movements their instincts prompt them to get to the ground, but in order to reach it they do not descend the body of the tree, neither do they cast off themselves precipitately; but, running to the side of the limbs they deliberately loosen their hold, and fall to the earth.

On reaching the ground the insects immediately bury themselves in the soil, burrowing by means of their broad and strong fore feet, which, like those of the mole, are admirably adapted for digging. In their descent into the earth they seem to follow the roots of plants, and are subsequently found attached to those which are most tender and succulent, perforating them with their beaks, and thus imbibing the vegetable juices which constitute their sole nourishment.

The grubs do not appear ordinarily to descend very deeply into the ground, but remain where roots are most abundant. The only alteration to which the insects are subject, during the long period of their subterranean confinement, is an increase in size, and the more complete development of the four small scale-like prominences on their backs, which represent and actually contain their future wings. Fig. 118, *a*, represents the full-grown larva.

When at length the time arrives for them to issue from the ground they come out in great numbers in the night, crawl up the trunks of trees, or upon any other object in their vicinity to which they can fasten themselves securely by their claws. After having rested awhile, they prepare to cast off their skins, which, in the mean time, have become dry and of an amber color. By repeated exertions, a longitudinal rent is made in the skin of the back, and through this the included Cicada pushes its head and body, and withdraws its wings and limbs from their separate cases, and, crawling to a little distance, it leaves its empty pupa-skin, apparently entire, still fastened to the tree as in fig. 118, *b*. At first the wing-covers are very small and opaque, but, being perfectly soft and flexible, they soon stretch out to their full dimensions, and in the course of a few hours the superfluous moisture of the body evaporates, and the insect becomes strong enough to fly.

During several successive nights the pupæ continue to issue from the earth; above fifteen hundred have been found to rise beneath a single apple tree, and in some places the whole surface of the soil, by their successive operations, has appeared as full of holes as a honey-comb. Within about a fortnight after their final transformation they begin to lay their eggs, and in the space of six weeks the whole generation becomes extinct.

They are subject to many accidents, and have many enemies, which contribute to diminish their numbers. Their eggs are eaten by birds; the young, when they first issue from the shell, are preyed upon by ants, which mount the trees to feed upon them, or destroy them when they are about to enter the ground. Blackbirds eat them when turned up by the plow in fields, and hogs are excessively fond of them, and, when suffered to go at large in the woods, root them up, and devour immense numbers just before the arrival of the period of their final transformation, when they are lodged immediately under the surface of the soil. It is stated that many perish in the egg state, by the rapid growth of the bark and wood, which closes the perforations and buries the eggs before they are hatched; and many, without doubt, are killed by their perilous descent from the trees.

Such are the general habits of this remarkable insect which now, and probably has for ages, passed seventeen years of its life hidden in the soil, and at stated periods has appeared for a short season of life above ground. A most elaborate account of the habits of this Cicada will be found in the First Missouri Report (1868), in which the important discovery is announced that there are races which complete their career in thirteen years; while no differences have been found between the Thirteen-year and Seventeen-year Cicadas, other than in the time of their appearing, yet some entomologists give them as distinct, and the Thirteen-year Cicadas are, for convenience called *C. tredecim*, Riley. This form is confined to more southerly localities than the other. While the Seventeen-year Cicada completes its round in that number of years, it is not due all over the country on the same year, but "the Locust years" differ in different States. In the Report above referred to, there are given the dates of the appearance of twenty-two different broods. For example. Brood XX which appeared in 1866, may be looked for in 1883 in Western New York, Western Pennsylvania and in Eastern Ohio. The brood XXI, is due in 1884 and at intervals of seventeen years thereafter, in parts of North Carolina and in Central Virginia. In 1885 a brood will appear in parts of New York and New England, in parts of Pennsylvania, Maryland, the District of Columbia, in Delaware, Virginia,

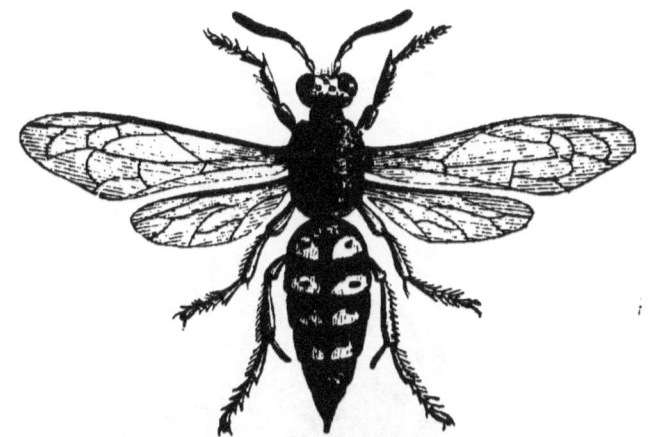

Fig. 122—THE DIGGER WASP (*Stizus grandis*).

in parts of Ohio, Michigan, Indiana and Kentucky. This brood has a record extending from 1715, since which date it has appeared at regular intervals of seventeen years, up to its last occurrence in 1868. In some cases two broods may lap over upon one another in the same locality.

DOES THE CICADA STING? There have been various accounts in the papers of injury from the sting of the Periodical Cicada. It has a beak by means of which it draws its nourishment from the branches of trees, and it may be that in careless handling of the insect, it has thrust its beak into the flesh. The most probable origin of these reports of stinging is due to the fact that a very large digger wasp (*Stizus grandis*) provides its nest with the Cicada, among other insects, as food for its young. The mother wasp stings her victims sufficiently to paralyze, but not to kill them, and takes them to her underground nest. This wasp is given of the real size in figure 122. It is possible that one may, in catching a Cicada, get a sting from this wasp, which had already captured it.

Insects Injurious to Small Fruits.

CURRANT AND GOOSEBERRY.

THE GOOSEBERRY SPAN-WORM.

(*Eufitchia ribearia*, Pack.)

This, which has been called the American Currant-moth, and sometimes merely "Currant-worm," was first described by Dr. Fitch, as *Abraxas ribearia*, referring it to the same genus with the European Gooseberry Moths. Later, Dr. Packard finding it to belong to a different genus, dedicated it to its distinguished discoverer, calling it *Eufitchia*. While it is found upon the Gooseberry, and Currant, it shows a decided preference for the former, and when the two are growing near one another, it will first attack the Gooseberry.

It may at once be distinguished from any other worm found either on Gooseberry or Currant, by its being what is popularly called a Measuring-worm. Figure 123, shows this larva, which, when full-grown, measures about an inch, and is of a light-yellow color, with lateral white lines, and numerous black spots and round dots. The head is white, with two black eye-like spots on the outer sides above, and two smaller ones beneath. The six true legs are black, and the four pro-legs yellow. It attains its growth about the middle of June, when it descends to the ground, and either burrows a little below the surface, or hides under any rubbish that may be lying there; but in neither case does it form any cocoon. Shortly after this it changes to a chrysalis, shown at the left-hand side in figure 123, of a shining mahogany color. In about fourteen days it bursts the pupa shell, and, early in July, appears as a moth, represented in figure 123, the upper one being the male, with feathered feelers, and the lower the female, in which these are simple. The

Fig. 123.—GOOSEBERRY SPAN-WORM (*Eufitchia ribearia*).

moth is of a pale nankin-yellow color, the wings shaded with faint dusky leaden-colored spots arranged so as not to present any definite pattern. The female lays her eggs on the branches and twigs of the bushes, hence the species is frequently carried in the egg state upon transplanted bushes from one neighborhood to another; which accounts for its sudden appearance in parts where it was before unknown. For there is but one brood of this insect in one year, and the eggs must consequently, like those of the Tent-worm of the Apple tree, be exposed, on the twigs and limbs to which they are attached, to all the heats of July and August without hatching out, and to all the frosts of December and January without freezing out. At length, when the proper time arrives, and the Gooseberry and Currant bushes are out in full leaf so as to afford plenty of food, the tiny tough little egg hatches out about the latter end of May, and in a little more than three weeks the worms attain their full larval development.

This Gooseberry Span-worm was first noticed near Chicago in 1862 or '63; and for two or three years afterwards it increased rapidly, so as in most gardens not to leave a single leaf on the Gooseberry, and in many instances to entirely strip the Currant bushes. It is quite common also in St. Louis and Jefferson Counties in Missouri, and has entirely stripped the Gooseberry bushes on many farms in these counties. Elsewhere in the Western States it is not by any means common; but in many localities in the East it has been a severe pest for a great number of years, especially in the States of New York and Pennsylvania. This is a native insect which no doubt formerly lived upon our several native species of Gooseberry. When cultivated Gooseberries were planted within their reach, they manifested a decided preference for these, and multiplied so rapidly as to become, in some localities, a serious pest to the fruit grower.

REMEDIES.—These worms, when disturbed, let themselves down from the bushes, and hang suspended by a web. This habit may be made useful in destroying them. If the bushes are shaken by means of a forked stick, while the worms are still young, these will at once let themselves down by their threads; the stick may be then passed along against the threads to draw the worms to the ground, where they may be crushed. Poultry may be used to capture the worms when they descend to the ground to transform. One of the most effective remedies for this, and all similar worms, is White Hellebore, used as described under Imported Currant-worm.

THE IMPORTED CURRANT-WORM.

(*Nematus ventricosus*, Ring.)

When the Currant-worm is mentioned in the Eastern States, this is most likely to be the insect referred to. It is the larva of a Saw-fly of the Order *Hymenoptera;* these are called False-caterpillars, as they never have less

Fig. 124.—THE IMPORTED CURRANT-WORM (*Nematus ventricosus*, Ring.)

than six, sometimes as many as eight pairs of pro-legs, while the True-caterpillars never have more than five pairs. The insect is a native of Europe, and is supposed to have been introduced into some of the nurseries at Rochester, N. Y., about the year 1857, though it appears to have been known in Canada at an earlier date. It is very abundant in the New England States, and in New York and Pennsylvania, and has extended to some of the Western States. The insect appears soon after the Currant and Gooseberry bushes put forth their leaves, and the eggs are laid upon the under surface of the lower leaves, along the principal veins. The eggs hatch in a week or ten days into a pale, twenty-legged Caterpiller, with a large dull whitish head. They soon become green, and acquire shining black spots on the body, and the head becomes black. The full-grown worms are about three-fourths of an inch long and are shown in various positions in figure 124, *a*; *b* gives the position of the black spots upon a magnified joint of the body. When they have completed their growth, they leave the bushes, and either hide just below the surface of the ground, or under any leaves that may be on the surface, spin a thin cocoon of brownish silk, within which they assume the pupa state.

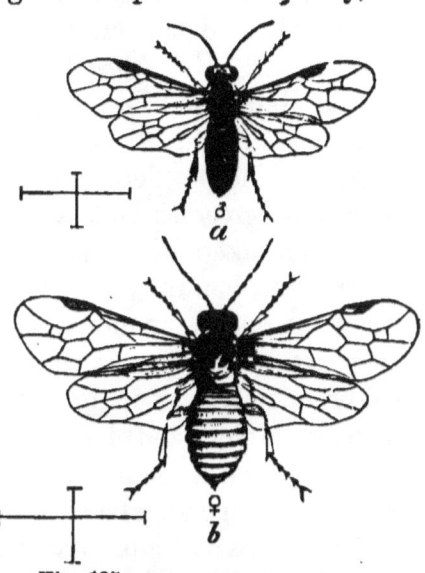

Fig. 125.—THE PERFECT INSECT.
a, Male; *b*, Female.

Late in June, or early in July, sometimes not until August, the perfect insects appear; a second crop of eggs is laid, and the same round is repeated; but this second brood does not issue from the pupa until the following spring. The perfect insect is shown in figure 125, *a* being the male, and *b* the female, the lines showing the actual size.

Those who receive Currant bushes from a distance, in order to avoid the introduction of this insect in the pupa state, should carefully wash the roots of the plants, and burn whatever may be washed from them.

REMEDIES.—When the worms are not checked, they soon strip both the Currant and Gooseberry bushes of their leaves, and the partly grown fruit shrivels and dies. The insect threatened to put an end to Currant culture in localities where it is an important crop, until an effective remedy was made known. By the prompt use of White Hellebore the insect may be subdued with but little trouble and the crop saved. Some papers speak of the use of "Hellebore," and it is necessary to specify White Hellebore (*Veratrum album*) which is an entirely different drug from the Black Hellebore (*Helleborus niger*). The powdered root, as sold at the drug stores, is of a light greenish-yellow color and excites violent sneezing when taken into the nostrils, hence care should be observed in handling it. The powder may be sprinkled upon the bushes by means of a tin sifter, but this is often attended by unpleasant sneezing, and is not so economical or effective as to apply it mixed with water. Place a heaping tablespoonful of the powder in a bowl or other dish holding a quart or more, gradually add boiling water, stirring to make sure that the powder is thoroughly wetted; then add more water, stirring until a quart, more or less, has been added. Turn this mixture into a pailful of cold water, stir well, and apply by the use of any garden syringe or hand engine, or a watering pot may be used. The object should be to wet every leaf, hence much force is not needed. In a few days, if any worms are found to have escaped, the application should be repeated; rarely are more than two doses needed. While White Hellebore is poisonous, no danger need be apprehended from the use of the fruit from bushes thus treated. The chances are that the rains will wash off any of the powder that may adhere to the clusters; but if any appreciable quantity should remain, the fruit would appear soiled, and be rejected on that account. The use of White Hellebore is so easy and so effective that none of the other applications that have been recommended need be noticed.

NATURAL ENEMIES.—There will be found in the IXth Missouri Report, a very full account of this Saw-fly, which states that it is attacked by several insect enemies, among which are the Placid Soldier Bug, and that there are at least two Ichneumon Flies that infest it.

THE NATIVE CURRANT WORM.

(Pristiphora grossulariæ, Walsh.)

This, like the Imported Currant-worm, is the larva of a Saw-fly, but of a different genus, distinguished by en-

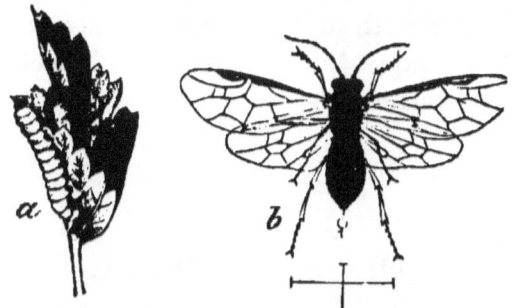

Fig. 126.—THE NATIVE CURRANT-WORM (*Pristiphora grossulariæ*, Walsh.)
a, Larva; *b*, Perfect Insect.

tomologists on account of the different veining of the wings. The larva is smaller than in the preceding, only half an inch long, and is of an uniform pale-green color, without any black dots. It does not go under-ground to make its cocoon, but always spins it among the twigs and leaves of the bushes. Figure 126 gives the larva of the natural size and the enlarged fly; the male and female being so nearly alike that separate figures are not needed. Unlike the preceding, the second brood issues the same autumn, and the eggs are laid upon the stems, where they pass the winter. Wherever this native insect occurs upon the cultivated Gooseberry and Currant, it may be subdued by the use of White Hellebore as recommended for the Imported Currant-worm.

THE CURRANT STALK-BORER.

(Ægeria tipuliformis, Linn.)

This is an imported insect and of the same genus as the Peach-borer. The moth lays her eggs singly near the buds, and the larvæ, when hatched, make their way directly to the pith, which they devour, forming a channel several inches in length. The stem, thus weakened, shows by the inferior size of its fruit that this insect is present, and it often breaks off at the affected part. The impoverished growth of the stems indicates the presence of this borer, and at the fall pruning, all such should be cut away and burned.

THE STRAWBERRY.

Among the insect enemies of the Strawberry, the common White Grub is probably one of the most destructive. This insect, which is injurious to so many different plants, is described in full, and suggestions for its suppression are given on page 33. Their injury to Strawberry plantations results mainly from bad management and the failure of the grower to use preventive measures. Good old pasture and meadow lands are frequently selected for Strawberry plantations, and sod is turned over, and as soon as sufficiently rotted, the plants are set out. In the meantime the grubs that were already in the ground, and perhaps of various ages from a few weeks to a year or two, have been fasting, or making an occasional meal of the half decayed grass-roots. Finding fresh Strawberry roots thrust before them, they commence a most vigorous attack upon such tender food. The planter is astonished to see his Strawberries disappear, and wonders where all the grubs could have come from in so short a time.

Now in regions where the White Grub abounds it is not safe to set out Strawberries on freshly inverted sod; but the land should be cultivàted at least two seasons in some crop requiring frequent hoeing and plowing, before using it for this purpose. Neither should the Strawberry plantation remain or be continued on the same piece of land for more that two or three years, if what is called the matted or bed system of cultivation is pursued; because the parent beetle soon learns that these weedy, little-disturbed plantations, are a safe place for her to deposit her eggs.

To avoid injury to Strawberry plantations by this insect, use land that has been occupied at least two years in some hoed crop, like corn, potatoes, or beans, and then set out a new one on fresh land as soon as the old plants begin to fail.

THE STRAWBERRY WORM.

(Emphytus maculatus, Norton.)

Among the various other kinds of insects injurious to the Strawberry there is perhaps none more destructive than that known as the "Strawberry Worm." This pest is a small, slender, pale-green worm, that attacks the leaves, eating large holes in them. When at all abundant it soon destroys the entire foliage, and of course prevents further growth of the plants. A. S. Fuller, in the "American Entomologist" says: A few years ago this

pest almost ruined the plants in my garden, but of late
it has not been very abundant, although it has not
entirely disappeared. This Strawberry Worm is the
larva of a small black fly, which has of late years
become abundant throughout the Northern States and
appears to be more destructive at the West than at the
East. The worms are of a yellowish-green color, a little
over half an inch long, and when feeding are usually curled up as in fig. 127. The parent fly (fig. 128), is black,
with two rows of whitish spots on the abdomen, and appears in the Northern States in May. The full-grown

Fig. 127.—STRAWBERRY-WORM (*Emphytus maculatus*, Norton).

Fig. 128.—FLY OF STRAWBERRY-WORM.

larvæ descend and enter the ground, remaining in the
pupa state until the following spring. Dusting the leaves
with lime, when they are wet with dew, or just after a
rain, is the best method of destroying the pest yet found.

STRAWBERRY LEAF-BEETLE.

(*Paria aterrima*, Oliv.)

Within a few years, in widely separated localities,
from Massachusetts to Missouri, a small brownish beetle
has been found attacking the leaves of the Strawberry
plants, doing much damage. The larva of this beetle is
white, with a yellowish head, and is about a fourth of an
inch long; it lives in the soil, feeding upon the roots of
the Strawberry. The beetle is only an eighth of an inch

long; the wing-covers are yellowish, and each usually has two black spots, the posterior one larger than the other. When numerous, it completely defoliates the plants. Paris Green or London Purple may be used when the fruit is off to destroy this insect in the manner described under Colorado Potato-beetle.

THE STRAWBERRY LEAF-ROLLER.

(*Anchylopera fragariæ*, Walsh and Riley.)

This insect, like the preceding, has been more troublesome at the West than at the East. It is the larva of a moth shown in figure 129, *a*, of the real size, while the moth, *b*, is enlarged, as shown by the lines below it. In feeding, it folds up the leaves by drawing their edges together with silken threads, and eats out the pulpy portions. There are two broods each year; the first one completing their transformations on the leaves about the first of July. The second brood of worms enter the ground, where they change into the pupæ, and remain in that state until the next spring. The habit the worms have of rolling themselves in the leaves, renders the application of any insecticide very difficult. It has been suggested to burn off the old leaves, after the fruit has been gathered, or to pass a heavy roller over the plants.

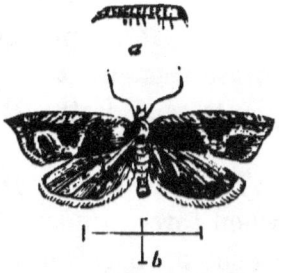

Fig. 129.—STRAWBERRY LEAF-ROLLER (*Anchylopera fragariæ*). *a*, Larva of real size; *b*, Moth enlarged.

THE STRAWBERRY CROWN-BORER.

(*Tyloderma fragariæ*, Riley.)

This enemy to the Strawberry grower has been more abundant in Canada and the Western States than elsewhere. The perfect insect is a small Snout-beetle, or Curculio, shown in figure 130, enlarged, the line giving the real size. The eggs are laid in the crowns of the plants, where the grubs destroy the embryo fruit stalks and leaves. The only remedy thus far suggested is, to plow up the infested plantations as soon as the fruit is gathered, while the young grubs are still in the crowns of the plants.

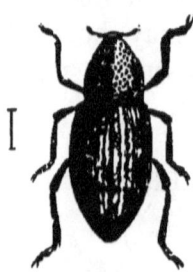

Fig. 130.—STRAWBERRY CROWN-BORER.

OTHER ENEMIES TO THE STRAWBERRY.

Sometimes a green fly or aphis, especially in light, loose soil, will attack the roots in large numbers. Dry ashes, or the use of the refuse dust of tobacco factories, applied close to these plants will destroy these insects; so would the use of tobacco-water.

The Red Spider is often injurious to the Strawberry, when forced under glass, and sometimes, in dry seasons, in the open ground. In either case, copious waterings, a thorough drenching of the leaves, is the best remedy.

THE BLACKBERRY.

Some ten years ago, the cultivators of the Blackberry in various parts of New Jersey noticed that the ends of the young growing canes in summer would occasionally curl, twist about, and often assume a singular, fasciated form, resulting in an entire check to their growth. The leaves on these infested shoots did not die and fall off, but merely curled up, sometimes assuming a deeper green than the healthy leaves on the same stalk. At the approach of winter the infested leaves remained firmly attached to the diseased stems, and all through the cold weather and far into the spring, these leaf-laden and diseased stems were a conspicuous object in many of the Blackberry plantations of the State.

If the infested shoots are examined in summer, thousands of minute insects of a pale-yellow color and covered with a powdery exudation will be found sucking the juices of the succulent stems and leaves, causing the crimping, curling, and twisting of these parts as described.

This parasite resembles somewhat an ordinary green-fly (*Aphis*) or plant-louse, but according to recent observations it is now known to belong to the closely allied Flea-lice family (*Psyllidæ*), distinguished from the plant-lice by a different veining of the wings, and by the antennæ being knobbed at the tip, like those of the butterfly, the knob usually terminating in two bristles. These insects jump as briskly as a flea, from which characteristic they derive their scientific name. The particular species in question was called by Prof. Riley the "Bramble Flea-louse (*Psylla rubi*)" in the "American Entomologist." It has increased very rapidly during the past half dozen years or more, and unless fruit-growers make a more vigorous fight than they yet have done, it will soon get the mastery of most Blackberry plantations. The only practical method as yet discovered for checking the ravages of this insect, is, to cut off the ends of the infested canes and burn them. This operation should always be performed either in the morning, or during cool wet weather, else many of the insects will escape, and at all times the severed shoots should be immediately dropped into bags and in them carried to the place where they are to be burned, and there emptied into the fire. If all having Blackberry bushes in their gardens would practice this method of destruction, this pest would soon cease to do much harm.

BLACKBERRY BORERS.

Several species of Borers infest the Blackberry: the most common one is the larva of a small, slender, red-necked beetle (*Oberea perspicillata*, Hald.), fig. 131. The small, legless grubs bore the pith of the canes, causing them to die prematurely, or so weakening them that they are broken down with the wind. As there are some fourteen or fifteen species of the *Oberea* now known, it may be that more than one species breed in the Blackberry. Thus far, however, I am not aware that they have been very injurious, but it would be well to gather all infested canes and burn them with their contents.

The Blackberry is subject to the attacks of several species of gall-insects. A fuzzy, prickly gall on the twigs is produced by a four-winged fly (*Diastrophus cuscutæformis* O. S.) Another species of the same genus (*Diastrophus nebulosus* O. S.) produces a large pithy gall on canes, but both of these gall-makers have very formidable parasitic enemies which keep them in check.

Fig. 131. BLACKBERRY BORER.

There are also a few leaf-eating beetles, slugs and caterpillars, that sometimes attack the Blackberry, but they are seldom sufficiently numerous or injurious to attract much attention. The larger species are readily destroyed by hand-gathering, and the smaller ones can usually be driven off by dusting the plants with lime.

The most formidable enemy however of both the Blackberry and Raspberry is the Orange-rust, a minute fungus (*Uredo ruborum*). It is perhaps more abundant on the Black-cap Raspberry (*Rubus occidentalis*) than on the ordinary varieties of the Blackberry; still it is sufficiently abundant and destructive to all to attract the attention of horticulturists throughout the country. I do not know of any remedy except to stamp out the disease by rooting up every affected plant and burning it. It may be that applications of lime, salt, or some similar substance would check the disease, and while these may be safely tried as preventive measures, the destroying of all infested plants should not be omitted.

THE RASPBERRY.

As the Raspberry is closely allied to the Blackberry, and belongs to the same genus, the diseases and insects infesting both do not materially differ. Some few species of insects seem to prefer the Raspberry, notably among which is what is called the the Red-necked Bupestris (*Agrilus ruficollis*, fig. 132), a small beetle that seems to be particularly fond of the red and black-cap varieties, but will occasionally attack the Blackberry. The larva bores into the canes in summer, causing large excrescences or galls (fig. 133), checking the flow of sap, and causing the death of the cane. This insect seems to be far more plentiful in the Western than Eastern States; but it is widely distributed, and every cultivator of the Raspberry may as well be on the lookout for it, and gather and burn all canes upon which galls of any kind are found.

Fig. 132. RED-NECKED BUPESTRIS.

Fig. 133.—GALLS IN RASPBERRY CANE.

THE SNOWY TREE-CRICKET.

(Œcanthus niveus, Harris.)

The Snowy Tree-Cricket, fig. 134, prefers the canes of the Raspberry for its eggs to the twigs of other shrubs or trees. It will, however, use the Grape, Willow, Peach, and other kinds, if Raspberries are not convenient. The long, slender eggs are deposited in a close compact row, an inch or more in length, each egg placed at a slight angle, and deep enough to reach the pith of the cane or twig in which it is set (fig. 135). This weakens the

Fig. 134.—SNOWY TREE-CRICKET (*Œcanthus niveus.*)

Fig. 135.— EGGS OF SNOWY TREE-CRICKET. *a*, Punctures in Stem; *b*, Eggs within Stem; *c*, Egg enlarged; *d*, Cap of Egg.

canes, and they are often broken off by the wind. This injury does not amount to much, but the perfect insect has a very bad habit of cutting off leaves in summer; and sometimes extends its mischievous work to the grape-vine, trimming off both leaves and fruit, working at night when perfectly safe from observation or molestation. One of my correspondents in Texas wrote me, a few years ago, that one of these pests would completely defoliate a young grape-vine in a single night, and he was a long time in discerning the successful nocturnal pruner, and when discovered he was at a loss how to circumvent it. Destroying the eggs is the only way thus far known of fighting this insect.

THE GRAPE-VINE.

THE HOG-CATERPILLAR OF THE VINE.

(*Chœrocampa pampinatrix*, Smith & Abbott.)

Of the large, solitary caterpillars that attack the Grape-vine, this is by far the most common and injurious in the Mississippi Valley. We have frequently found the egg of this insect glued singly to the underside of a leaf. It is 0.05 inch in diameter, perfectly round, and of a uniform

Fig. 136.—HOG-CATERPILLAR OF THE VINE (*Chœrocampa pampinatrix*.)

delicate yellowish-green color. The young worm which hatches from it, is pale-green, with a long straight horn at its tail; and after feeding from four to five weeks it acquires its full growth, when it presents the appearance of figure 136, the horn having become comparatively shorter and acquired a posterior curve.

This worm is readily distinguished from other grape-feeding species by having the third and fourth rings immensely swollen, while the first and second rings are quite small and retractile. It is from this peculiar appearance of the fore part of the body, which strikingly suggests the fat cheeks and shoulders and small head of some breeds of swine, that it may best be known as the Hog-caterpillar of the vine. The color of this worm when full grown is pea-green, and it is wrinkled transversely and covered with numerous pale-yellow dots, placed in irregular transverse rows. An oblique cream-colored lateral band, bordered below with a darker green and most distinct on the middle segments, connects with a cream-colored subdorsal line, which is bordered above with darker green, and which extends from the head to the horn at the tail. There are five and often six somewhat pale-yellow triangular patches along the back, each containing a lozenge-shaped lilac-colored spot. The head is small, with yellow granulations, and four perpendicular yellow lines, and the stigmata or spiracles are orange-brown.

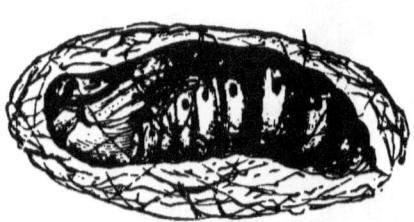

Fig. 137.—CHRYSALIS OF HOG-CATERPILLAR.

When about to transform, the color of this worm usually changes to a pinkish-brown, the darker parts being of a beautiful mixture of crimson and brown. Previous to this change of color Mr. J. A. Lintner has observed the worm to pass its mouth over the entire surface of its body, even to the tip of its horn, covering it with a coating of apparently glutinous matter—the operation lasting about two hours. Before transforming into the pupa or chrysalis state, it descends from the vine, and within some fallen leaf or under any other rubbish that may be lying on the ground, forms a mesh of strong brown silk, within which it soon changes to a chrysalis (fig. 137) of a pale, warm yellow, speckled and spotted with brown, but characterized chiefly by the conspicuous dark brown spiracles and broad brown incisures of the three larger abdominal segments.

The moth (fig. 138) which in time bursts from this chrysalis, has the body and front wings of a fleshy-gray, marked and shaded with olive-green, while the hind wings are of a deep rust-color, with a small shade of gray near their inner angle.

This insect is in northerly regions one-brooded, but towards the south two-brooded, the first worms appearing in the latitude of St. Louis, during June and July, and giving out the moths about two weeks after they become chrysalids, or from the middle of July to the first

Fig. 138.—MOTH OF HOG-CATERPILLAR.

of August. The second brood of worms are full grown in September and, passing the winter in the chrysalis state, give out the moths the following May. On one occasion we found at South Pass, Ill., a worm half grown and still feeding as late as October 20th, a circumstance which would lead to the belief, that at points where the winters are mild they may even hibernate in the larva state.

This worm is a most voracious feeder, and a single one will sometimes strip a small vine of its leaves in a few nights. According to Harris it does not even confine its attacks to the leaves, but in its progress from leaf to leaf, stops at every cluster of fruit, and either from stupidity, or disappointment, nips off the stalks of the half-grown grapes and allows them to fall to the ground untasted. It is fortunate for the grape-grower therefore that Nature has furnished the ready means to prevent its ever becoming excessively numerous, for we have never known it to swarm in very great numbers. The obvious reason is, that it is so freely attacked by a small parasitic Ichneumon-fly—belonging to a genus (*Microgaster*) exceedingly numerous in species—that three out of every four worms we meet with will generally be found to be thus victimized. The eggs of the parasite are deposited within the body of the worm, while it is yet young, and the young maggots hatching from them feed on the fatty parts of their victim. After the last moult of a worm that has been thus attacked, numerous little heads may be seen gradually pushing through different parts of its body; and as soon as they have worked themselves so far out that they are held only by the last joint of the body, they commence forming their small snow-white cocoons, which stand on end, pushes open a little lid which it had previously cut with its jaws, and soars away to fulfil its mission. It is one of those remarkable and not easily explained facts, which often confront the student of Nature, that, while one of these Hog-caterpillars in its normal and healthy condition may be starved to death in two or three days, another that is writhing with its body full of parasites will live without food for as many weeks. Indeed we have known one to rest for three weeks without food in a semi-paralyzed condition, and after the parasitic flies had all escaped from their cocoons, it would rouse itself and make a desperate effort to regain strength by nibbling at a leaf which was offered to it. But all worms thus attacked succumb in the end, and the grape-grower should let alone all such as are found to be covered with white cocoons, and not, as has been often done, destroy them under the false impression that the cocoons are the eggs of the worm. The cocoons of a parasite are shown upon another large larva, on page 88; figure 59.

THE ACHEMON SPHINX.

(*Philampelus achemon*, Drury.)

This is another large Grape-vine-feeding insect, belonging to the great *Sphinx* family, and which may be popularly known as the Achemon Sphinx. It has been found in almost every State where the Grape is cultivated, and also in Canada. It feeds on the American Woodbine or Virginia Creeper (*Ampelopsis quinquefolia*) with as much relish as on the Grape-vine, and seems to show no preference for any of the different varieties of the latter. It is, however, worthy of remark, that both its food-plants belong to the same Botanical Family.

The full grown worm or larva is usually found during the latter part of August and fore part of September. It measures about three and one-half inches when crawling, which operation is effected by a series of sudden jerks. The third segment is the largest, the second but half its size, and the first still smaller, and when at rest the two last mentioned segments are partly withdrawn into the third. The young larva is green, with a long slender reddish horn rising from the eleventh segment and curving over the back, and though we have found full grown specimens that were equally as green as the younger ones, they more generally assume a pale-straw or reddish-brown color, and the long recurved horn is invariably replaced by a highly polished lenticular tubercle. It is often of a pale-straw color which deepens at the sides and finally merges into a rich vandyke-brown. The worm is covered more or less with minute spots which are dark on the back but light and annulated at the sides, while there are from six to eight transverse wrinkles on all but the thoracic and caudal segments.

The color of the worm, when about to transform, is often of a most beautiful pink or crimson. The chrysalis is formed within a smooth cavity under ground. It is of a dark shiny mahogany-brown color, shagreened or roughened, especially at the anterior edge of the segments on the back.

Unlike the Hog-caterpillar of the Vine, this insect is everywhere single-brooded, the chrysalis remaining in the ground through the fall, winter, and spring months, and producing the moth towards the latter part of June.

The moth is of a brown-gray color, handsomely variegated with light-brown, and with dark deep brown spots. The hind wings are pink with a dark shade across the middle, still darker spots below this shade, and a broad gray border behind.

We have never found any parasite attacking this species, but its solitary habit and large size make it a conspicuous object, and it is easily controlled by hand, whenever it becomes unduly numerous upon the Grape-vine.

THE SATELLITE SPHINX.

(Philampelus satellitia, Linn.)

Like the Achemon Sphinx, this insect occurs in almost every State in the Union. It also bears a strong resemblance to the former species, and likewise feeds upon the Virginia Creeper (*Ampelopsis*), as well as upon the Grape-vine; but the worm may be distinguished by having five cream-colored spots each side, instead of six, and by the spots themselves being less scalloped.

In the latitude of St. Louis, this worm is found full grown throughout the month of September, and a few specimens may even be found as late as the last of October. The eggs of this species, as of all other Hawk-moths (*Sphinx* family) known to us, are glued singly to the leaf of the plant which is to furnish the future worm with food. When first hatched, and for some time after-

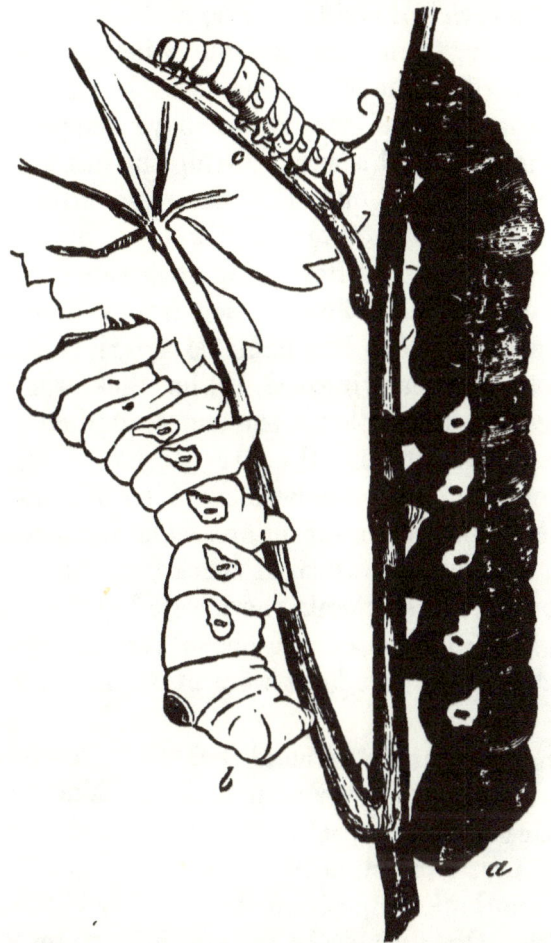

Fig. 139.—CATERPILLAR OF SATELLITE SPHINX (*Philampelus satellitia*, Linn.)
a, Mature Larva; *b*, at rest; *c*, Young Larva.

wards, the larva is green, with a tinge of pink along the sides, and with an immensely long straight pink horn at the tail. This horn begins to shorten, and finally curls round like a dog's tail, as at figure 139, *c*. As the worm grows older it changes to a reddish-brown, and by the third moult it entirely loses the horn.

When full grown, it measures nearly four inches in length, and when crawling appears as figure 139, *a*. It crawls by a series of sudden jerks, and will often fling its head savagely from side to side when alarmed. Dr. Morris describes the mature larva as being green, with six side patches; but though we have happened across many specimens of this worm during the last seven years, we never once found one that was green after the third moult; nor do we believe that there are ever any more than five full-sized yellow spots each side, even in the young individuals. The specimen from which our figure was made, occurred at Hermann, Missouri, in Mr. George Husmann's former vineyard. The back was pinkish, inclining to flesh-color; the sides gradually became darker and darker, and the five patches on segments 6 to 10 inclusive, were cream-yellow with a black annulation, and shaped as in our figure. On segments 2, 3, 4, 5 and 6, were numerous small black dots, but on each of the following five segments there were but two such dots. A pale longitudinal line ran above the yellow patches, and the head and first joint were uniformly dull reddish-brown.

The most common general color of the full grown worm is a rich velvety vinous-brown. When at rest, it draws back the fore part of the body, and retracts the head and first two joints into the third (fig. 139, *b*), and in this motionless position it no doubt manages to escape from the clutches of many a hungry insectivorous bird.

When about to transform, the larva of our Satellite Sphinx enters a short distance into the ground, and soon works off its caterpillar-skin and becomes a chrysalis of a deep chestnut-brown. The moth (fig. 140) makes its appearance in June of the following year, though it has been known to issue the same year that it had existed as larva. In this last event, it doubtless becomes barren, like others under similar circumstances. The colors of

Fig. 140.—THE SATELLITE SPHINX (*Philampelus satellitia*, Linn.)

the moth are light olive-gray, variegated as in the figure with dark olive-green. The worms are easily subdued by hand-picking.

THE ABBOT SPHINX.

(*Thyreus Abbotii*, Swainson.)

This is another of the large Grape-feeding insects, occurring on the cultivated and indigenous vines and on the Virginia Creeper, and having, in the full grown larva state, a polished tubercle instead of a horn at the tail. Its habitat is given by Dr. Clemens, as New York, Pennsylvania, Georgia, Massachusetts, and Ohio; but though not so common as the Sphinx Moths already described, yet it is often met with both in Illinois and

Fig. 141.—THE ABBOT SPHINX (*Thyreus Abbotii*, Swain.)
Larva and Moth.

Missouri. The larva which is represented in the upper part of figure 141 varies considerably in appearance. Indeed, the ground-color seems to depend in a measure on the sex, for Dr. Morris describes this larva as reddish-brown with numerous patches of light-green, and expressly states that the female is of a uniform reddish-brown, with an interrupted dark-brown dorsal line and transverse striæ lines. We have reared two individuals which came to their growth about the last of July, at which time they were both without a vestige of green. The ground-color was dirty yellowish, especially at the sides. Each segment was marked transversely with six or seven slightly impressed fine black lines, and longitudinally with wider non-impressed dark-brown patches, alternating with each other, and giving the worm a checkered appearance. These patches become more dense along the subdorsal region, where they form two irregular dark lines, which on the thoracic segments become single, with a similar line between them. There was also a dark stigmatal line with a lighter shade above it, and a dark stripe running obliquely downwards from the posterior to the anterior portion of each segment. The belly was yellow with a tinge of pink between the prolegs, and the shiny tubercle at the tail was black, with a yellowish ring around the base. The head, which is characteristically marked, and by which this worm can always be distinguished from its allies—no matter what the ground-color of the body may be—is slightly roughened and dark, with a lighter broad band each side, and a central mark down the middle which often takes the form of an **X**. This worm does not assume the common Sphinx attitude of holding up the head, but rests stretched at full length, though if disturbed it will throw its head from side to side, thereby producing a crepitating noise.

The chrysalis is formed in a superficial cell on the ground; its surface is black and roughened by confluent punctures, but between the joints it is smooth and inclines to brown; the head-case is broad and rounded, and the tongue-case is level with the breast; the tail terminates in a rough flattened wedge-shaped point, which gives out extremely small thorns from the end.

The Moth (figure 141,) appears in the following March or April, there being but one brood each year. It is of a dull chocolate or grayish-brown color, the front wings becoming lighter beyond the middle, and being variegated with dark brown as in the figure; the hind wings are sulphur-yellow, with a broad dark-brown border breaking into a series of short lines on a flesh-colored ground, near the body. The wings are deeply scalloped, especially the front ones, and the body is furnished with lateral tufts. When at rest, the abdomen is curiously curved up in the air.

THE BLUE CATERPILLARS OF THE VINE.

Besides the large Sphinx caterpillars, described and figured on the preceding pages, every grape-grower must have observed certain so-called "Blue Caterpillars," which, though far from being uncommon, are yet very rarely sufficiently numerous to cause alarm, though in some few cases they have been known to strip certain vines. There are three distinct species of these blue caterpillars, which bear a sufficient resemblance to one another, to cause them to be easily confounded. The first and by far the most common in the West, is the larva of

THE EIGHT-SPOTTED FORESTER.

(*Alypia octomaculata*, Fabr.)

This larva (fig. 142, *a*), may often be found in the latitude of St. Louis as early as the beginning of May, and more abundantly in June, while scattering individuals (probably of a second brood) are even met with, but half-grown, in the month of September. The young larvæ are whitish with transverse lines, the colors not contrasting so strongly as in the full-grown specimens, though the black spots are more conspicuous. They feed beneath the leaves and can let themselves down by a web,

The full-grown larva often conceals itself within a folded leaf. It is of the form of our figure, and is marked transversely with white and black lines, each segment having about eight light and eight dark ones. The bluish appearance of this caterpillar is owing to an optical phenomenon from the contrast of these white and black stripes. The head and the shield on the first segment are of a shiny bright deep orange color, marked with black dots, and there is a prominent transverse orange-red band, faint on segment 2 and 3, conspicuous on 4 and 11, and uniform in the middle of each of the other segments. In the middle segments of the body each orange band contains eight black elevated spots, each spot giving rise to a white hair. These spots are arranged as in the enlarged section shown in the engraving (fig. 142, *b*), namely, four on each side, as follows: the upper one on the anterior border of the orange band, the second on its posterior border, the third just above spiracles

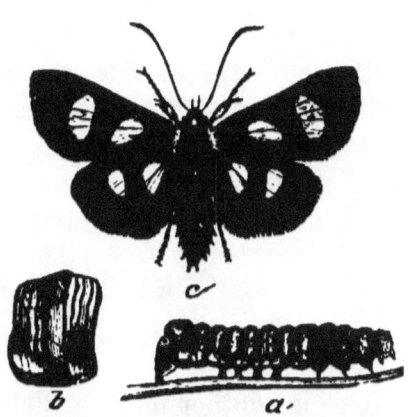

Fig. 142.—EIGHT-SPOTTED FORESTER (*Alypia octomaculata*, Fabr.) *a*, Larva; *b*, Section; *c*, Moth.

on its anterior border—each of the three interrupting one of the transverse black lines—and the fourth, which is smaller, just behind the spiracles. The venter is black, slightly variegated with bluish-white, and with the orange band extending on the legless segment. The legs are black, and the false legs have two black spots on an orange ground, at their outer base, but the characteristic feature, which especially distinguishes it from the other two species, is a lateral white wavy band—obsolete on the thoracic segments, and most conspicuous on 10 and 11—running just below the spiracles, and interrupted by the transverse orange band.

This larva transforms to chrysalis within a very slight cocoon formed without silk, upon, or just below, the surface of the earth, and issues soon after, as a very beautiful moth of a deep blue-black color, with orange shanks, yellow shoulder-pieces, each of the front wings with two large light yellow spots, and each of the hind wings with two white ones. Figure 142, *c*, represents the female, and the male differs from her in having the wing spots larger, and in having a conspicuous white mark along the top of his narrower abdomen.

We have on one or two occasions known vines to be partly defoliated by this species, but never knew it to be quite so destructive as it often is in some Eastern localities. In New York City the vines in the yards are often completely stripped of their foliage through the agency of this and related caterpillars.

THE BEAUTIFUL WOOD NYMPH.

(*Eudryas grata*, Fabr).

Here is another moth which surpasses in real beauty, though not in high contrast, the species just described. The front wings are milk-white, broadly bordered and marked on their margins with rusty-brown, the band on the outer margin being shaded on the inner side with olive-green, and marked towards the edge with a slender wavy white line: under surface yellow, with two dusky spots near the middle. The hind wings are nankin-yellow, with a deep-brown border, which does not extend to the outer angle, and which also contains a wavy white line: under surface yellow, with a single black spot.

Surely these two moths are as unlike in general appearance as two moths well can be; and yet their caterpillars bear such a close resemblance to each other, and both feed upon the Grape-vine. The larva of the Beautiful Wood Nymph is, in fact, so very similar to that of the Eight-spotted Forester, that it is entirely unnecessary to figure it. It differs more especially from that species by invariably lacking the white patches along the sides; the hairs arising from the black spots are less conspicuous, while the hump on the eleventh segment is somewhat more prominent. The light parts of the body have really a slight bluish tint, and in specimens which we have found, we have only noticed six transverse black stripes to each segment. This larva, when at rest, depresses the head and raises the third and fourth segments, Sphinx-fashion. It is found on the vines in Missouri as early as May and as late as September, and it devours all portions of the leaf, even to the midrib. It descends to the ground, and, without making any cocoon, transforms to a chrysalis, which is dark colored, rough, with the tip of the abdomen obtusely conical, ending in four tubercles, the pair above, long and truncate, those below broad and short. Some of them give out the moth the same summer, but most of them pass the winter and do not issue as moths until the following spring.

THE PEARL WOOD NYMPH.

(*Eudryas unio*, Hübner).

This little moth is also closely allied to, and much resembling the preceding species. It is smaller, and differs from the Beautiful Wood Nymph in having the outer border of the front wings paler and of a tawny color, with the inner edge wavy instead of straight; and in that of the hind wings being less distinct, more double, and extending to the outer angle.

The larva is said by Dr. Fitch to so much resemble that of the preceding species that " we as yet know not whether there are any marks whereby they can be distinguished from each other." The moth is more common in the West than its larger ally, and though we have never bred it from the larva, yet we have often met with a worm which, for various reasons, we take to be this species. It never grows to be quite so large as the other, and may readily be distinguished by its more decided bluish cast; by having but four light and four dark stripes to each segment, by having no orange band across the middle segments, and by the spots, with the exception of two on the back placed in the middle light band, being almost obsolete. The head, shield on first segment, hump on the 11th, and a band on the 12th, are orange, spotted with black. Venter orange, becoming dusky towards head; feet and legs also orange, with blackish extremities, and with spots on their outside at base.

This worm works for the most part in the terminal buds of the vine, drawing the leaves together by a weak silken thread, and cankering them. It forms a simple earthen cocoon, or frequently bores into a piece of old wood, and changes to chrysalis, which averages but 0.36-inch in length: this chrysalis is reddish-brown, covered on the back with rows of very minute teeth, with the tip of the abdomen truncated, and terminating above in a thick blunt spine each side.

From the above accounts, we hope our readers will have no difficulty in distinguishing between these three blue caterpillars of the Grape-vine.

REMEDIES.—The larvæ of the two Wood Nymphs have a fondness for boring into old pieces of wood, to transform to the chrysalis state, and Mr. T. B. Ashton, of White Creek, N. Y., found that they would even bore into corn cobs for this purpose in preference to entering the ground, wherever such cobs were accessible. The Eight-spotted Forester, on the contrary, has no such habit, and while the only mode of combating it is to pick the larvæ off and burn them, the Wood Nymphs may be more easily subdued by scattering a few corn-cobs under the vines in the summer—to be raked up and burned in the winter. It has been suggested that many of these moths might be destroyed by exposing poisoned molasses or syrup at the time of their appearance in spring. White Hellebore as described under Currant and Gooseberry would no doubt be efficacious, and good results may be expected to follow the use of Pyrethrum, or Persian Insect Powder.

THE GRAPE LEAF-FOLDER.

(*Desmia maculalis*, West.)

This has long been known to depredate on the leaves of the Grape-vine in many widely separated parts of North America. It is not uncommon in Canada West, and is found in the extreme southern parts of Georgia. It appears to be far more injurious, however, in the intermediate country, or between latitude thirty-five and forty degrees, than in any other sections, and in Southern Illinois and Central Missouri proves more or less injurious every year. It belongs to the same family as our notorious Clover-worm, which attacks our clover stacks and mows.

This genus is characterized by the elbowed or knotted appearance of the male antennæ, in contrast with the smooth, thread-like female antennæ; the maxillary palpi are not visible, while the compressed and feathery labial palpi are recurved against the eyes, and reach almost to their summit; the body extends beyond the hind wings. The moth of the Grape Leaf-folder is a very pretty little thing, expanding on an average almost an inch, with a length of body of about one-third of an inch. It is conspicuously marked, and the sexes differ sufficiently to have given rise to two names, the female having been named *Botys bicolor*. The color is black, with an opalescent reflection, and the under surface differs only from the upper in being less bright; all the wings are bordered with white. The front wings of both sexes are each furnished with two white spots; but while in the male (fig. 143, 4), there is but one large spot on the hind wings, in the female (fig. 143, 5), this spot is invariably more or less constricted in the middle, especially above, and is often entirely divided into two distinct spots. The body of the male has but one distinct transverse band, and a longitudinal white dash at its extremity superiorly, while that of the female has two white bands. The antennæ,

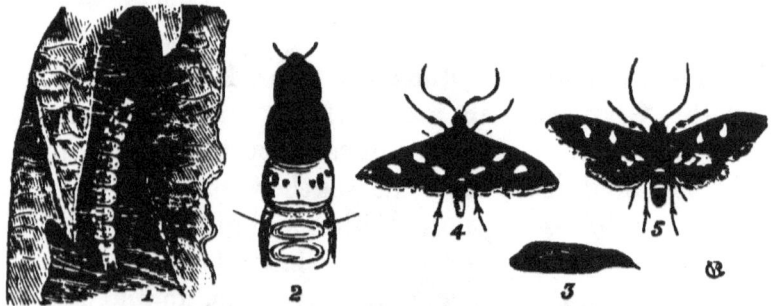

Fig. 143.—GRAPE LEAF-FOLDER (*Desmia maculalis*, West.)
1, Worm; 2, Head, etc., enlarged; 3, Chrysalis; 4, Male; 5, Female.

as already stated, are still more characteristic, those of the male being elbowed and thickened near the middle, while those of the female are simple and thread-like.

There are two broods in this latitude—and probably three farther south—during the year; the first moths appearing in June, the second in August, and the worms produced from these last hibernating in the chrysalis state. The eggs are scattered in small patches over the vines, and the worms are found of all sizes at the same time. These last change to chrysalids in twenty-four to thirty days from hatching, and give forth the moths in about a week afterwards.

The worm (fig. 143, 1), folds rather than rolls the leaf,

by fastening two portions together by its silken threads; and for this reason, in contradistinction to the many leaf-rollers, may be popularly known as the "Grape Leaf-folder." It is of a glass-green color, and very active, wriggling, jumping and jerking either way at every touch. The head and thoracic segments are marked as at figure 143, 2. If let alone these worms will soon defoliate a vine, and the best method of destroying them is by crushing suddenly within the leaf, with both hands. To prevent their appearance, however, requires far less trouble. The chrysalis is formed within the fold of the leaf, and by going over the vineyard in October, or any time before the leaves fall, and carefully plucking and destroying all those that are folded and crumpled, the supply for the following year will be cut off. This should be done collectively to be positively effectual, for the utmost vigilance will avail but little if one is surrounded with slovenly neighbors.

We believe this insect shows no preference for any particular kind of grape-vine, having found it on all the cultivated, as well as the wild varieties. Its natural enemies are Spiders, Wasps, and a small *Tachina* fly, which attacks it in the larva state, and a small clay-yellow beetle is supposed to attack it.

THE COMMON YELLOW BEAR.

(*Spilosoma Virginica*, Fabr.)

This is one of the most common North American insects. The moth, which is very generally dubbed "the Miller," frequently flies into our rooms at night.

Though the moth is so common, how few persons ever think of it as the parent of that frequent and most troublesome of caterpillars, which Harris has so aptly termed the Yellow Bear. These caterpillars are quite frequently found on the Grape-vine, and when about one-fourth grown bear a considerable resemblance to the mature larva of the Grape-vine Plume. They seldom appear, however, until that species has disappeared, and may always be distinguished from it by their semi-gregarious habit at this time of their life, and by living exposed on the leaf (generally the underside) instead of forming a retreat within which to hide themselves, as does the Plume.

The Yellow Bear is found of all sizes from June to October; and though quite fond of the Vine, is by no means confined to that plant. It is, in fact, a very general feeder, being found on a great variety of herbaceous plants, both wild and cultivated, as butternut, lilac, beans, peas, convolvulus, corn, currant, gooseberry, cotton, sunflower, plantain, smart-weed, verbenas, geraniums, and almost any other plant with soft, tender leaves. These caterpillars are indeed so indifferent as to their diet, that we have actually known one to subsist entirely, from the time it cast its last skin till it spun up, on dead bodies of the Camel Cricket (*Mantis Carolina*).

When young they are invariably bluish-white, but when full grown they may be found either of a pale cream-color, yellow, light brown, or very dark-brown, the different colors often appearing in the same brood of worms, as we have proved by experiment. Yellow is the most common color, and in all the varieties the venter is dark, and there is a characteristic longitudinal black line, more or less interrupted, along each side of the body, and a transverse line of the same color (sometimes faint) between the joints; the head and feet are ochre-yellow, and the hairs spring from dark yellow warts, of which there are ten on each joint, those on joint 1 being scarcely distinguishable, and those on joint 12 coalescing. There are two broods of these worms each year, the broods intermixing, and the last passing the winter in the chrysalis state. The chrysalis is formed in a trivial cocoon, constructed almost entirely of the caterpillar's hairs, which, though held in position by a few very fine silken threads, are fastened together mainly by the interlocking of their minute barbs, and the manner in which the caterpillar interweaves them.

The best time to destroy these worms is soon after they hatch from their little round yellow eggs, which are deposited in clusters; for, as already intimated, they then feed together.

THE GRAPE-VINE PLUME.

(Pterophorus periscelidactylus, Fitch.)

Just about the time that the third bunch of grapes, on a given shoot, is developing, many of the leaves, and especially those at the extremity of the shoot, are found fastened together more or less closely, but generally so as to form a hollow ball. These leaves are fastened by a fine white silk, and upon opening the mass and separating the leaves, one or two caterpillars will generally be found in the retreat. We say one or two, because the retreat made by the smallest of the Blue-caterpillars of the Vine, namely, the larva of the Pearl Wood Nymph, so closely resembles that of the Grape-vine Plume under consideration, that until the leaves are separated it is almost impossible to tell which larva will be found. Both occur at the same time of year. In an ordinary season they do not draw together the tips of the shoots until after the third bunch of grapes is formed, and in devouring the terminal bud and leaves, they do little more than assist the vineyardist in the pruning which he would soon have to give. They act, indeed, as Nature's pruning-knives. But the severe frost which generally kills the first buds, so retards the growth of the vines that the worms come out in full force before the third bunch has fully formed, and this bunch is consequently included in the fold made by these worms, and destroyed.

The larva of the Grape-vine Plume invariably hatches very soon after the leaves begin to expand; and though it is very generally called the Leaf-folder, it must not be confounded with the true Leaf-folder, described on page 231, and which does its principal damage later in the season. At first the larva of our Plume is smooth and almost destitute of hairs, but after each moult the hairs become more perceptible, and when full grown the larva appears as at figure 144, *a*, the hairs arising from a transverse row of warts, each joint having four above and six below the breathing pores (see fig. 144, *e*). After feeding for about three weeks our little worm fastens itself securely by the hind legs to the underside of some leaf or other object, and, casting its hairy skin, transforms to the pupa state. The pupa (fig. 144, *b*), with the lower part of the three or four terminal joints attached to a little silk previously spun by the worm, hangs at a slant of about forty degrees. It is of peculiar and characteristic form, being ridged and angular, with numerous projections, and having remnants of the larval warts; it is obliquely truncated at the head, but is chiefly distinguished by two com-

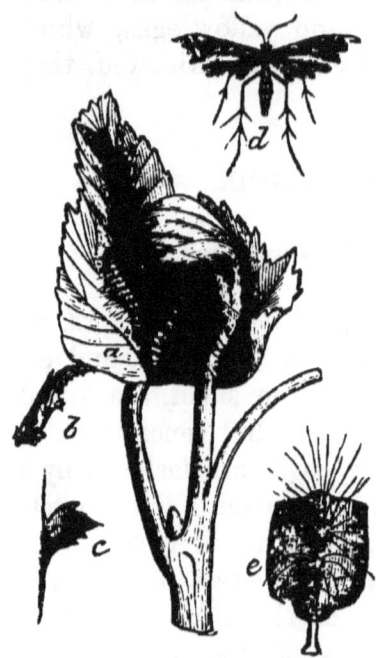

Fig. 144.—GRAPE-VINE PLUME (*Pterophorus periscelidactylus*, Fitch.)
a, Larva; *b*, Pupa; *c*, Horn; *d*, Moth; *e*, Hairs.

pressed sharp-pointed horns, one of which is enlarged at figure 144, *c*, projecting from the middle of the back: it measures, on an average, rather more than one-third of an inch, and varies in color from light green with darker green shadings, to pale straw-color with light-brown shadings.

The moth (fig. 144, *d*), escapes from this pupa in about one week, and, like all the species belonging to the genus, it has a very active and impetuous flight, and rests with the wings closed and stretched at right angles from the body, so as to recall the letter T. It is of a tawny yellow color, the front wings marked with white and dark brown as in the figure, the hind wings appearing like burnished copper, and the legs being alternately banded with white and tawny yellow.

All the moths of the family (*Alucitidæ*) to which it belongs have the wings split up into narrow feather-like lobes, and for this reason they have very appropriately been called Plumes in popular language. In the genus *Pterophorus* the front wings are divided into two, and the hind wings into three lobes. In this country, a somewhat larger species (*P. carduidactylus*, Riley) occurs on the Thistle, and though bearing a close resemblance to the Grape-vine Plume in color and markings, yet differs very remarkably in the larva and pupa states.

From analogy we infer that there are two broods of these worms each year, and that the last brood passes the winter in the moth state. We have, however, never noticed any second appearance of them, and whether this is from the fact that the vines are covered with a denser foliage in the summer than in the spring, or whether there is really but one brood, are points in the history of our little Plume which yet have to be settled by further observation.

On account of its spinning habit this insect is easily kept in check by hand picking.

THE GRAPE-BERRY MOTH.

(*Penthina vitivorana*, Packard.)

The Grape-berry Moth is an illustration of the well-known fact that an insect may suddenly appear in many different parts of the country where it had not been known before, for previous to 1878 no account of it had been published, and it was entirely unknown to science. It had however been noticed in several localities in Ohio, Illinois, and Missouri, for three or four years, but never so abundant as in 1878. In that year it was common in Missouri, in Illinois, and ruined about fifty per cent. of the grapes around Cleveland, Ohio. It has also appeared in Pennsylvania, and may appear at any time where grapes are grown.

Its natural history may be given as follows: About the 1st of July, the grapes that are attacked by the worms begin to show a discolored spot at the point where the worm entered, (fig. 145, *c*). Upon opening such a grape, the inmate, which is at this time very small and white, with a cinnamon-colored head, will be found at the end of a winding channel. It continues to feed on the pulp of the fruit, and upon reaching the seeds, generally eats out their interior. As it matures it becomes darker, being either of an olive-green or dark-brown color, with a honey-yellow head, and if one grape is not sufficient, it fastens the already ruined grape to an adjoining one, by means of silken threads, and proceeds to burrow in it as it did in the first. When full grown it presents the appearance of figure 145, *b*, and is exceedingly active. As soon as the grape is touched the worm will wriggle out of it, and rapidly let itself to the ground, by means of its ever-ready silken thread, unless care be taken to prevent its so doing. The cocoon is often formed on the leaves of the vine, in a manner essentially characteristic. After covering a given spot with silk, the worm cuts out a clean oval flap, leaving it hinged on one side, and, rolling this flap over, fastens it to the leaf, and thus forms for itself a cozy little house. One of these cocoons is represented at figure 146, *b*, and though the cut is sometimes less regular than shown in the figure, it is undoubtedly the normal habit of the insect to make just such a cocoon as represented. Sometimes, however, it cuts two crescent-shaped slits, and, rolling up the two pieces, fastens them up in the middle as shown at figure 147. And frequently it rolls over a piece of the edge of the leaf, in the manner commonly adopted by leaf-rolling larvæ,

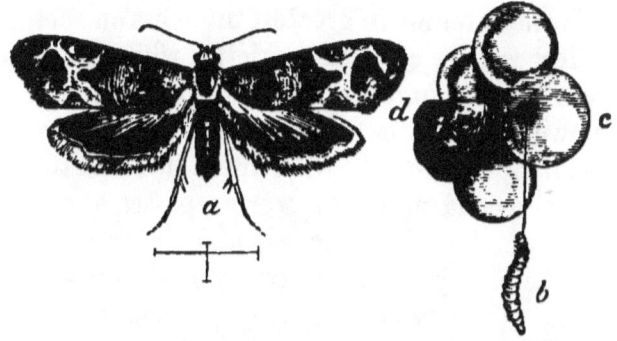

Fig. 145.—GRAPE BERRY-MOTH (*Penthina vitivorana*, Packard.)
a, Moth; *b*, Larva; *c*, Punctured Berry; *d*, Shrunken Berry.

while we have had them spin up in a silk handkerchief, where they made no cut at all.

In two days after completing the cocoon, the worm changes to a chrysalis. In this state (fig. 146, *a*), it measures about one-fifth of an inch, and is quite variable in color, being generally of a honey-yellow, with a green shade on the abdomen. In about ten days after this last change takes place, the chrysalis works itself almost entirely out of the cocoon, and the little moth represented at figure 145, *a*, makes its escape.

The first moths appear in Southern Illinois and Central Missouri about the 1st of August, and as the worms are found in the grapes during the months of August and September, or even later, and as Mr. Read has kept the cocoons through the greater part of the winter, there is every reason to believe that a second brood of worms is generated from these moths, and that the second brood of worms, as is the case of the Codling-moth of the apple, passes the winter in the cocoon, and produces the moth the following spring, in time to lay the eggs on the grapes while they are forming.

Fig. 146.
GRAPE BERRY-MOTH.
a, Pupa; *b*, Chrysalis.

This worm is found in greatest numbers on such grapes as the Herbemont, or those varieties which have tender skins, and close, compact bunches; though it has also been known to occur on almost every variety grown. As already stated, there can be little doubt but that the greater part of the second brood of worms passes the winter in the cocoon on the fallen leaves; and, in such an event, many of them may be destroyed by raking up and burning the leaves at any time during the winter. The berries attacked by the worm may easily be detected, providing there is no "grape rot" in the vineyard, either by a discolored spot as shown at figure 145, *c*, or by the entire discoloration and shrinking of the berry, as is shown at

Fig. 147.—CHYSALIS.

figure 145, *d*. When the vineyard is attacked by the "rot," the wormy berries are not so easily distinguished, as they bear a close resemblance to the rotting ones. All fallen berries should be picked up and destroyed.

THE GRAPE-VINE FLEA-BEETLE.

(*Graptodera* [formerly *Haltica*] *chalybea*, Illig.)

Of the numerous insect enemies with which our grape-growers have to deal, this occupies a prominent place.

The beetles which have hibernated begin their destructive work in the spring as soon as the buds commence to swell, and it is at this early period that the greatest dam-

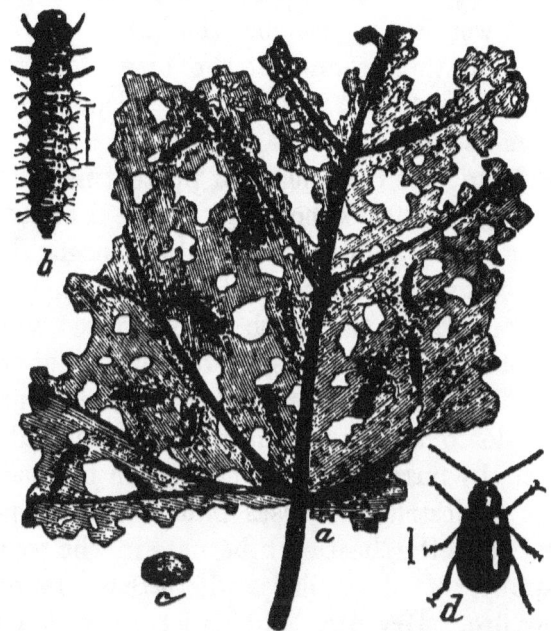

Fig. 148.—GRAPE-VINE FLEA-BEETLE (*Graptodera chalybea*, Illig.)
a, Young Larvæ on Leaf; *b*, Larva, enlarged; *c*, Chrysalis; *d*, Beetle.

age is done by the beetles boring into and feeding on said buds. Later in the season the beetles feed upon the leaves, and upon these, in the month of May, the female lays her small orange-colored eggs in clusters. These soon hatch, and the young dark-colored larvæ riddle the leaf as shown in figure 148, *a*, or when very numerous completely devouring it, leaving only the largest ribs.

In about a month the full-grown larvæ (fig. 148, *b*), descend into the ground, where each forms a small earthen cell (fig. 148, *c*), and changes to a dull-yellowish pupa of the shape normally assumed in this family. The perfect beetle issues about three weeks later, from the middle of June to the middle of July, and again begins to eat the leaves, but the damage done is trifling compared with that done in early spring. So far as we have observed there is but one annual generation, but it is probable that in the more Southern States there will be two. As soon as cold weather approaches the beetles retire under fallen leaves in the ground, at the base of trees, under loose bark, in houses, in short, in any place which offers shelter from the cold.

In considering the best means of preventing the injuries of this insect, it must be borne in mind, that, according to our observations, the female beetle deposits her eggs by preference on the leaves of the wild grape vines, as the larvæ are rarely met with in cultivated vineyards. It is against the perfect beetle, therfore, that we must direct our efforts at destruction, and while it is undoubtedly desirable to keep the vineyard clear of rubbish in winter time, by burning wherever fire can be used safely, this means of destruction loses much of its importance by the fact that the beetles hibernate in the woods and in any number of other places where they cannot be destroyed by fire. Dry lime and hellebore, which may be used to advantage against the larvæ, have proved useless against the beetle, while lye and soapsuds cannot be used strong enough to kill it without injurious effects upon the plant. Tin pans or pails with some liquid at the bottom have been used to advantage for collecting the early beetles, which could be knocked into them, and we have repeatedly advised for this and other insects that infest the grape-vine, which fall to the ground upon disturbance, the use of sheets along the trellis to catch them. Unless repeatedly shaken from such sheets into vessels containing liquid, the beetles will of course soon escape.

The wonderful efficacy of kerosene in destroying insect life has long been known. It was used with excellent effect in shallow tin pans, or on stretched sheets of cloth, for the destructive locust of the West.

Mr. I. O. Howard, Assistant Entomologist to the Department of Agriculture, employed it successfully on sheets against the Grape-vine Flea-beetle, finding it so satisfactory that he did not hesitate to recommend it in the following terms:

"Take two pieces of common cotton sheeting, each being two yards long and half as wide; fasten sticks across the ends of each piece to keep the cloth open, and then drench with kerosene. Give the sheets thus prepared to two persons, each having hold of the rods at the opposite ends of the sheets. Then let these persons pass one sheet on either side of the vine, being careful to unite the cloth around the base of the vine; then let a third person give the stake to which the vine is attached a sharp blow with a heavy stick. Such a blow will in nearly every case jar the beetles into the sheets, where the kerosene kills them almost instantly.

"This process, after a little experience, can be performed almost as rapidly as the persons employed can walk from one vine to another. The expense necessary is very trifling, and boys can do the work quite as well as men. Warm bright afternoons are the proper times for this work to be done, and it should be performed faithfully every sunny day until the vines are out of danger."

Until something is discovered, which, blown or syringed on the buds, will keep off the beetles, this method of Mr. Howard's of dealing with the insect, will remain the best yet known.

THE SPOTTED PELIDNOTA.

(*Pelidnota punctata*, Linnæus.)

This is the largest and most conspicuous beetle that attacks the foliage of the Grape-vine, and in the beetle state it seems to subsist entirely on the leaves of this plant, and of the closely allied Virginia Creeper. Though some years it becomes so abundant as to badly riddle the foliage of our vineyards, yet such instances are exceptional; and it usually occurs in such small numbers, and

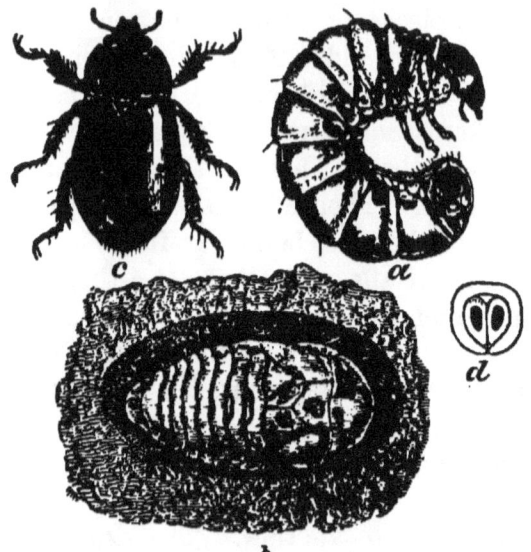

Fig. 149.—THE SPOTTED PELIDNOTA (*Pelidnota punctata*, Linn.)
a, Grub; *b*, Pupa; *c*, Beetle; *d*, Markings.

is so large and clumsy, that it can not be considered a very redoubtable enemy.

Its larva has, for a number of years, been known to feed on the decaying roots of different trees. It is a large clumsy grub (fig. 149, *a*), bearing a close resemblance to the common White Grub of our meadows, and differs from that species principally in being less wrinkled, and in having the chitinous covering (or skin, so-called) more polished and of a purer white color, and in the distinct heart-shaped swelling above the vent (fig. 149, *d*). Towards the latter part of June we have found this larva in abundance, in company with the pupa (fig. 149, *b*), in rotten stumps and roots of the Pear. In preparing for the pupa state, the larva forms a rather unsubstantial cocoon of its own excrement, mixed with the surrounding wood. The pupa state lasts but from eight to ten days, and the beetle (fig. 149, *c*), is found on our vines during the months of July, August, and September. It is not yet known how long a time is required for the development of the larva, but from analogy we may infer that the insect lives in that state upwards of three years.

This beetle was named about a century ago by Linnæus, who met with a specimen in the magnificent collection of shells and insects belonging to Queen Louise Ulrica of Sweden. It occurs throughout the States and Upper Canada, and is even met with in the West Indies. It flies and feeds by day. The wing-covers are of a slightly metallic clay-yellow color, with three distinct black spots on each, and the wings themselves are dark-brown inclining to black; the thorax is usually a little darker than the wing-covers, with one spot each side; the abdomen beneath, and legs, are of a bronzed-green. It is easily kept in check by hand-picking.

THE ROSE-BUG, OR ROSE-CHAFER.

(*Macrodactylus subspinosus*, Fabr.)

This insect does its injurious work in the beetle state. The larva develops under ground. The following account is condensed from the standard work of Harris. In arranging insects according to the plants to which they are injurious, it is difficult to decide where to place this; if we take into account the pecuniary loss it causes, perhaps the grape-grower is the greatest sufferer, and it is accordingly placed among the insects especially injurious to the Grape:

"The prevalence of this insect on the Rose, and its annual appearance coinciding with the blossoming of that flower, have gained for it the popular name by which it is here known. For some time after they were first noticed, Rose-bugs appeared to be confined to their favorite, the blossoms of the Rose; but within forty years they have greatly increased in number, have attacked at random various kinds of plants in swarms, and have become notorious for their extensive and deplorable ravages. The Grape-vine, in particular, the Cherry, Plum, and Apple trees, have annually suffered by their depredations; many other fruit trees and shrubs, garden vegetables and corn, and even the trees of the forest, and grass of the fields, have been laid under contribution by these indiscriminate feeders, by whom leaves, flowers, and fruits, are alike consumed. The unexpected arrival of these insects in swarms at the first coming, and their sudden disappearance at the close of their career, are remarkable facts in their history. They come forth from the ground during the second week in June, or about the time of the blossoming of the Damask Rose, and remain from thirty to forty days. At the end of this period the males perish, while the females enter the earth, lay their eggs, return to the surface, and, after lingering a few days, die also.

"The eggs laid by each female are about thirty in number, and are deposited from one to four inches beneath the surface of the soil; they are nearly globular, whitish, and are one-thirtieth of an inch in diameter, and are hatched twenty days after they are laid. The young larvæ begin to feed on such tender roots as are within their reach. They attain their full size in autumn, being then nearly three-quarters of an inch long, and about an eighth of an inch in diameter. In October they descend below the reach of frost, and pass the winter in a torpid state. In the spring they approach toward the surface, and each one forms for itself a little cell, of an oval shape. Within this cell the grub is transformed to a pupa during the month of May. During the month of June, the included beetle bursts open its earthen cell, and digs its way to the surface of the ground. Thus the various changes, from the egg to the full development of the perfected beetle, are completed within the space of one year."

The beetle is given of its real size, about seven-twentieths of an inch in length, in figure 150; its body is entirely covered with a very short and close ashen-yellow down; its legs are of a pale-red color, while the joints of the very long feet are tipped with black.

Fig. 150. ROSE-BUG.

REMEDIES.—Such being the metamorphoses and habits of the Rose-bugs, it is evident we cannot attack them in the egg, the grub, or the pupa state. When they have issued from their subterranean retreats, and have congregated upon our vines, trees, and other vegetable productions, in the complete enjoyment of their propensities, we must unite our efforts to seize and crush the invaders. They must indeed be crushed, scalded, or burned, to deprive them of life, for they are not affected by any of the applications usually found destructive to other insects. Experience has proved the utility of gathering them by hand, or of shaking them or brushing them from the plants into tin vessels containing a little water. They should be collected daily, especially in early morning, when they are torpid, and burned or scalded. If a film of kerosene is floated upon the water in the vessels in which they are caught, it will help to prevent their escape.

THE GRAPE PHYLLOXERA.

(*Phylloxera vastatrix*, Planchon.)

This minute insect, which has caused such devastations in the vineyards of Europe, is a native of this country, where its destructive work was known long before the cause of it was discovered. The life history of the Phylloxera has been worked out by Prof. Riley in his Missouri Reports, especially in the Sixth, from which the following account is condensed.

The insect presents itself under several different forms, all of which belong to two types. One of these is the Leaf-gall type (*gallicola*, R.), and the other is found upon the roots of the vine (*radicicola*, R.).

First, as to the Leaf-gall Type (*Gallicola*.)— The gall or excrescence produced by this is a fleshy swelling of the under side of the leaf, more or less wrinkled and hairy, with a corresponding depression of the upper side, the margin of the cup being fuzzy, and drawn together so as to form a fringed mouth. It is usually cup-shaped, but some times greatly elongated or purse-shaped (figure 151, *a, b*).

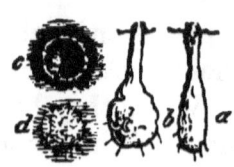

Fig. 151.
a and *b*, Elongated Galls; *c* and *d*, upper and under side of Abortive Galls.

Soon after the first vine-leaves that put out in the spring have fully expanded, a few scattering galls may be found, mostly on the lower leaves, nearest the ground. These vernal galls are usually large (of the size of an ordinary pea,) and the normal green is often blushed with rose where exposed to the light of the sun. On carefully opening one of them (fig. 152, *d*), we shall find the mother-louse diligently at work surrounding herself with pale-yellow eggs of an elongate oval form, scarcely .01-inch long, and not quite half as thick (fig. 152, *c*). She is about .04-inch long, generally spherical in shape, of a dull-orange color, and looks not unlike an immature seed of the common purslane. At times, by the elongation of the abdomen, she is more or less perfectly pear-shaped. Her members are all dusky, and so short, compared to her swollen body, that she appears very clumsy, and undoubtedly would be outside of her gall, which she never has occasion to quit, and which

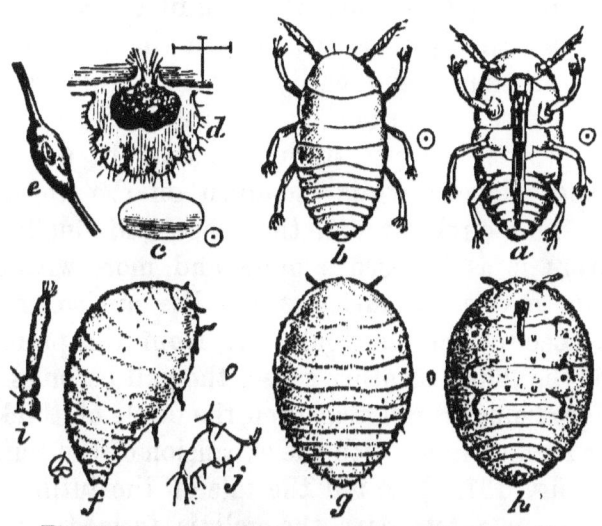

Fig. 152.—GRAPE PHYLLOXERA—LEAF-GALL TYPE.

a, b, Newly-hatched Larva, ventral and dorsal view; *c,* Egg; *d,* Section of Gall, *e,* Swelling of Tendril; *f, g, h,* Mother Gall-louse—lateral, dorsal and ventral views; *i,* her Antenna; *j,* her two-jointed Tarsus. Natural sizes indicated at sides by small circles.

serves her alike as dwelling house and coffin. More carefully examined, her skin is seen to be shagreened or minutely granulated and furnished with rows of minute hairs. The eggs begin to hatch, when six or eight days old, into active little oval, six-footed beings, which differ from their mother in their brighter yellow color and more perfect legs and antennæ, the tarsi being furnished with long, pliant hairs, terminating in a more or less distinct globule. In hatching, the egg splits longitudinally from the anterior end, and the young louse, whose pale-yellow is in strong contrast with the more dusky color of the egg-shell, escapes in the course of two minutes. Issuing from the mouth of the gall, these young lice scatter over the vine, most of them finding their way to the tender terminal leaves, where they settle in the downy bed which these leaves affords, and commence pumping up and appropriating the sap. The tongue-sheath is blunt and heavy, but the tongue proper—consisting of three brown, elastic, and wiry filaments, which, united, make so fine a thread as scarcely to be visible with the strongest microscope—is sharp, and easily run into the leaf. Its puncture causes a curious change in the tissues of the leaf, the growth being so stimulated that the under side bulges and thickens, while the down on the upper side increases in a circle around the louse, and finally hides and covers it as it recedes more and more within the deepening cavity. Sometimes the lice are so crowded that two occupy the same gall. If, from the premature death of the louse, or other cause, the gall becomes abortive before being completed, then the circle of thickened down or fuzz enlarges with the expansion of the leaf, and remains (fig. 151, *c*), to tell the tale of the futile effort. Otherwise, in a few days the gall is formed, and the inheld louse, which, while eating its way into house and home, was also growing apace, begins a parthenogenetic maternity by the deposition of fertile eggs, as her immediate parent had done before. She increases in bulk with pregnancy, and one egg follows another in quick succession, until the gall is crowded. The mother dies and shrivels, and the young, as they hatch, issue and found new galls. This process continues during the summer until the fifth or sixth generation. Every egg brings forth a fertile female, which soon becomes wonderfully prolific. The number of eggs found in a single gall averages about two hundred; yet it will sometimes reach as many as five hundred. Even supposing there are but five generations during the year, and taking the lowest of the above figures, the immense prolificacy of the species becomes manifest. As summer advances, they frequently become prodigiously multiplied, completely covering the leaves with their galls, when they appear as in figure 153. The lice also settle on the tendrils, leaf-stalks, and tender branches, where they also form knots and rounded excrescences (figure 152, e), much resembling those made on the roots. In such a case, the vine loses its leaves prematurely. Usually, however, the

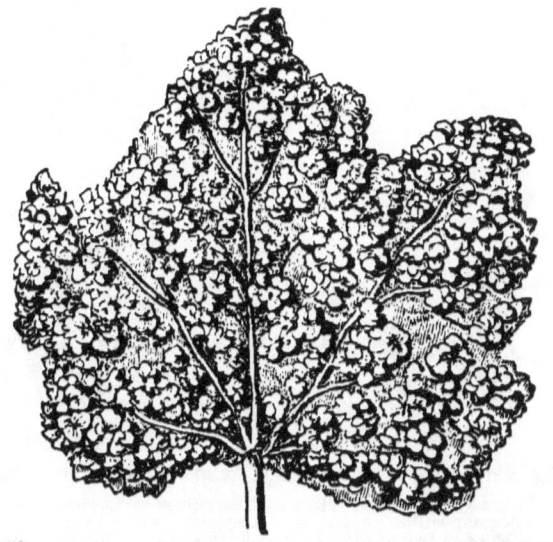

Fig. 153.—LEAF OF THE GRAPE-VINE WITH PHYLLOXERA GALL.

natural enemies of the louse seriously reduce its numbers by the time the vine ceases its growth in the fall, and the few remaining lice, finding no more succulent and suitable leaves, seek the roots. Thus, by the end of September, the galls are mostly deserted, and those which are left are almost always infested with mildew, and eventually turn brown and decay. On the roots, the young lice attach themselves singly or in little groups, and thus hibernate. The male gall-louse has never been seen, and there is every reason to believe that he has no existence. Nor does the female ever acquire wings. It is but a transient summer state, not at all essential to the perpetuation of the species, and does, compared with the other type, but trifling damage.

As already indicated, the autumnal individuals of *gallicola* descend to the roots, and there hibernate. There is every reason to believe also that, throughout the sum-

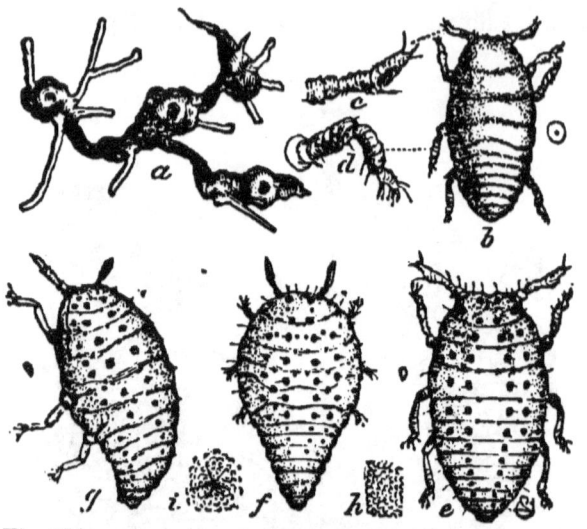

Fig. 154.—GRAPE PHYLLOXERA, ROOT-INHABITING TYPE.

a, Roots of Clinton vine, showing relation of Swellings to Leaf-galls, and power of resisting decomposition; *b*, Larva as it appears when hibernating; *c*, *d*, Antenna and Leg of same; *e*, *f*, *g*, Forms of more mature Lice; *h*, Granulations of Skin; *i*, Tubercle.

mer, some of the young lice hatched in the galls are passing on to the roots; as, considering their size, they are great travellers, and show a strong disposition to drop, their natural lightness enabling them thus to reach the earth with ease and safety. At all events, we know from experiment, that the young *gallicola*, if confined to vines on which they do not normally form galls, will, in the middle of summer, make themselves perfectly at home on the roots.

THE ROOT-INHABITING TYPE (*Radicicola*).—We have seen that, in all probability, *gallicola* exists only in the wingless, shagreened, non-tubercled, fecund female form. *Radicicola*, however, presents itself in two principal forms. The newly hatched larvæ of this type are undistinguishable, in all essential characters, from those hatched in the galls; but in due time they shed the smooth larval skin, and acquire raised warts or tubercles which at once distinguish them from *gallicola*. In the development from this point the two forms are separable with sufficient ease: one (A) of a more dingy greenish-yellow, with more swollen fore-body, and more tapering abdomen; the other (B) of a brighter yellow, with the lateral outline more perfectly oval, and with the abdomen more truncated at tip.

The first or mother form (fig. 154, *f*, *g*), is the analogue of *gallicola*, as it never acquires wings, and is occupied, from adolescence till death, with the laying of eggs, which are less numerous and somewhat larger than those found in the galls. We have counted in the spring as many as two hundred and sixty-five eggs in a cluster, and all evidently from one mother, who was yet very plump, and still occupied in laying. As a rule, however, they are less numerous. With pregnancy this form becomes quite tumid and more or less pyriform, and is content to remain with scarcely any motion in the more secluded parts of the roots, such as creases, sutures, and depressions, which the knots afford. The skin is distinctly shagreened (fig. 154, *h*,) as in *gallicola*. The warts, though usually quite visible with a good lens, are at other times more or less obsolete, especially on the abdomen.

The second or more oval form (fig. 154, *e*), is destined to become winged. Its tubercles, when once acquired, are always conspicuous; it is more active than the other, and its eyes increase rather than diminish in complexity with age. From the time it is one-third grown, the little dusky wing-pads may be discovered, though less conspicuous than in the pupa state, which is soon after

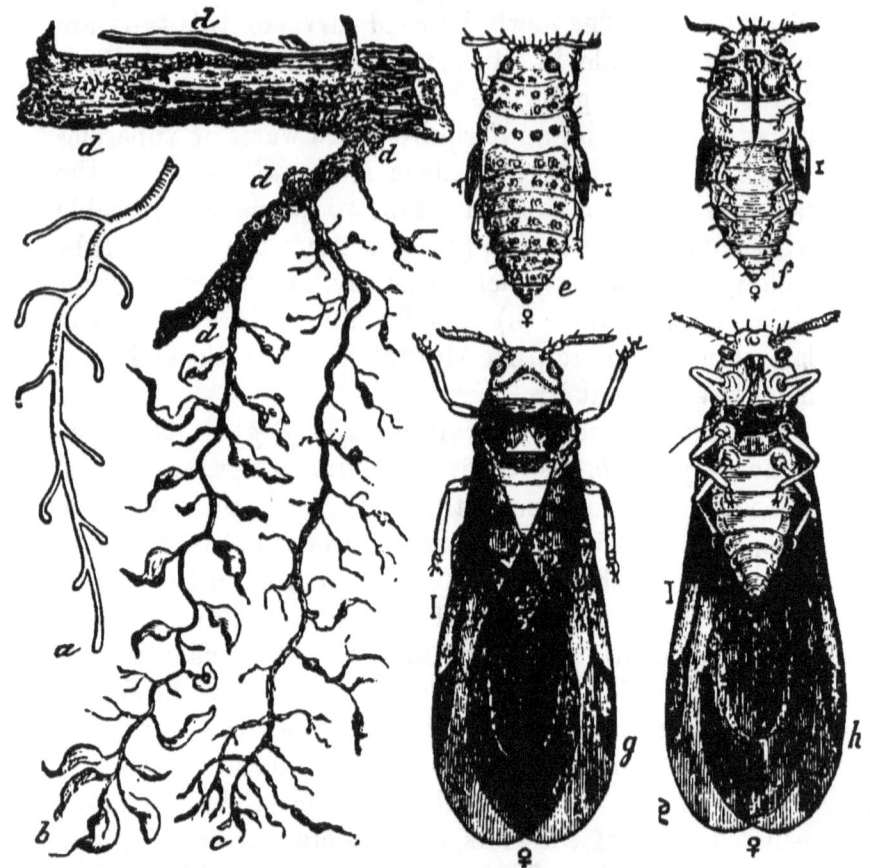

Fig. 155.—GRAPE PHYLLOXERA, ROOT-INHABITING TYPE.
a, Shows a healthy root; *b*, one on which the lice are working, representing the knots and swellings caused by their punctures; *c*, a Root that has been deserted by them, and where the rootlets have commenced to decay; *d, d, d*, show how the lice are found on the larger roots; *e*, female pupa, dorsal view; *f*, same, ventral view; *g*, winged female, dorsal view; *h*, same, ventral view.

assumed. The pupæ (fig. 155, *e*, *f*), are still more active, and, after feeding a short time, they make their way to the light of day, crawl over the ground and over the vines, and finally shed their last skin and assume the winged state, which is shown in figure 155, *g* and *h*. In this last moult the tubercled skin splits on the back, and is soon worked off, the body in the winged insect having neither tubercles nor granulations.

These winged insects are most abundant in August and September, but may be found as early as the first of July, and until the vines cease growing in the fall. The majority of them are females, with the abdomen large, and more or less elongate. From two to five eggs may invariably be found in the abdomen of these, and are easily seen when the insect is held between the light, or mounted in balsam or glycerine. A certain proportion have an entirely different shaped and smaller body, the abdomen being short, contracted, and terminating in a fleshy and dusky protuberance; the limbs stouter, and the wings proportionately larger and stouter.

This form has been looked upon as the male by myself, Planchon, Lichtenstein and others. Yet we have never succeeded in witnessing it performing the functions of a male, nor has any one else that we are aware of. The males in all plant-lice are quite rare, and, in the great majority of species, unknown.

As fall advances the winged individuals become more and more scarce, and as winter sets in, only eggs, newly-hatched larvæ, and a few wing-less, egg-bearing mothers are seen. These last die and disappear during the winter, which is mostly passed in the larva state, with here and there a few eggs. The larvæ thus hibernating (fig. 154, *b*), become dingy, with the body and limbs more shagreened and the claws less perfect than when first hatched; and, of thousands examined, all bear the same appearance, and all are furnished with strong suckers. As soon as the ground thaws and the sap starts in the spring, these young lice work off their winter coat, and, growing apace, commence to deposit eggs.

At this season of the year, with the exuberant juices of

the plant, the swellings on the roots are large and succulent, and the lice plump to repletion. One generation of the mother form (A) follows another—fertility increasing with the increasing heat and luxuriance of summer—until at least the third or fourth has been reached before the winged form (B) makes its appearance in the latter part of June or early in July.

Since (in 1870) the absolute identity of these two types was proved, by showing that the gall-lice become root-lice, the fact has been repeatedly substantiated by different observers. (In 1873 galls were obtained on the leaves of a Clinton vine from the root-inhabiting type, thus establishing the identity of the two types).

THE MORE MANIFEST AND EXTERNAL EFFECTS OF THE PHYLLOXERA DISEASE.—The result which follows the puncture of the root-louse is an abnormal swelling, differing in form according to the particular part and texture of the root. These swellings, which are generally commenced at the tip of the rootlets, eventually rot, and the lice forsake them and betake themselves to fresh ones —the living tissue being necessary to the existence of this as of all plant-lice. The decay affects the parts adjacent to the swellings, and on the more fibrous roots cuts off the supply of sap to all parts beyond. As these last decompose, the lice congregate on the larger ones, until at last the root system literally wastes away. The appearance of the root fibres before and after they have been attacked by the insect, is shown in figure 155, *a*, *b*, *c*.

During the first year of the attack there are scarcely any outward manifestations of disease, though the fibrous roots, if examined, will be found covered with nodosities, particularly in the latter part of the growing season. The disease is then in its incipient stage. The second year all these fibrous roots vanish, and the lice not only prevent the formation of new ones, but, as just stated, settle on the larger roots, which they injure, and which become disorganized and rot. At this stage the outward symptoms of the disease first become manifest, in a sickly, yellowish appearance of the leaf and a reduced growth of cane. As the roots continue to decay, these symptoms become more acute, until by about the third year the vine dies. Such is the course of the malady on the European vine (*V. vinifera*), when circumstances are favorable to the increase of the pest. When the vine is about dying it is generally impossible to discover the cause of the death, the lice, which had been so numerous the first and second years of invasion, having left for fresh pasturage.

MODE OF SPREADING.—The gall-lice can only spread by travelling, when newly hatched, from one vine to another; and if this slow mode of progression were the only one which the species is capable of, the disease would be comparatively harmless. The root-lice, however, not only travel under ground along the interlocking roots of adjacent vines, but crawl actively over the surface of the ground, or wing their way from vine to vine, and from vineyard to vineyard. Doubts have been repeatedly expressed by European writers as to the power of such a delicate and frail-winged fly to traverse the air to any great distance.

But there is abundant evidence as to their power of flight; they have been caught in spider-webs in Europe, and have been captured on sheets of paper prepared with bird-lime, and suspended in an infested vineyard, and there is no doubt that they can sustain flight for a considerable time under favorable conditions, and, with the assistance of the wind, they may be wafted to great distances. These winged females are much more numerous in the fall of the year than has been supposed. Whereever they settle, the few eggs which each carries are sufficient to perpetuate the species, and thus spread the disease, which, in the fullest sense, may be called contagious.

SUSCEPTIBILITY OF DIFFERENT VINES TO THE DISEASE.—As a means of coping with the Phylloxera disease, a knowledge of the relative susceptibility of different varieties to the attacks and injuries of the insect is of paramount importance. As is often the case with injurious insects, the Phylloxera shows a preference for and thrives best on certain species, and even discriminates between varieties; or, what amounts to the same thing, practically, some varieties resist its attacks, and enjoy a relative immunity from its injuries. It may be stated that there is a relation between the susceptibility of the vine and the character of its roots—the slow-growing, more tender-wooded, and consequently more tender-rooted varieties succumbing most readily; the more vigorous growers resisting best. The European Vine (*Vitis vinifera*), in its many varieties, is little affected by the leaf-inhabiting type, but it succumbs in a few years to the root-lice. Varieties of the Northern Fox-grape (*V. Labrusca*) vary much; some, like the Concord and others, resist well, while others, like the Catawba, suffer severely. Varieties derived from *V. æstivalis* and *V. cordifolia* are nearly exempt from the root-form, but some of them have the leaves much attacked by the gall-type. The Southern Fox-grape (*V. vulpina*) is entirely free from Phylloxera in any form.

REMEDIES AND PREVENTIVES.—Thus far, the only practicable method of combating the insect when established upon the root, is by drowning it by irrigating the soil. In Europe, the method largely adopted is to graft their vines upon varieties, the roots of which are Phylloxera proof; for this purpose American varieties have been sent to Europe in immense numbers, as cuttings and as rooted plants. An enterprising grape-growing firm has even established nurseries in Europe for the production of vines that resist the Phylloxera.

THE GRAPE LEAF-HOPPER.

(Tettigonia vitis, Harris.)

In many parts of the country, if one passes through a vineyard during July or August, he will be annoyed by the clouds of a small insect which, as it flies, appears as if it were of a dirty white color. These insects are generally known as "Thrips," a name belonging to a different genus, and which should be superseded by Leaf-hopper. The insect belongs to the Order *Hemiptera,* or true Bugs. It is the *Tettigonia vitis* of Harris (though some authors place it in *Erythroneura*), who thus describes it: "In its perfect state it measures one-tenth of inch in length. It is of a pale-yellow color; there are two little red lines on the head. The back part of the thorax, the scutel, the base of the wing-covers, and a broad band across their middle are scarlet; the tips of the wing-covers are blackish, and there are some little red lines between the broad band and the tips. The head is crescent-shaped above, and the eyelets are situated just below the ridge of the front." The insects appear upon the underside of the leaves in June, but are not much noticed, as they do not have their wings until later. They pass their larvæ state quietly, sucking at the juices of the leaves, which they penetrate with their beaks, though if disturbed at this time, they leap from leaf to leaf in a lively manner. They undergo all their changes on the leaves, and their empty skins may be found on the underside of the leaves, or upon the ground beneath the vine, in great numbers. The insect probably hibernates in the perfect state, hidden in the rubbish and in tufts of grass. When present in great numbers, they rob the vine of its proper nutri-

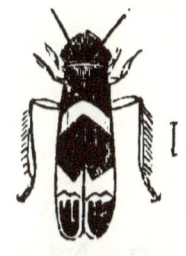

Fig. 156.—GRAPE LEAF-HOPPER.

ment, and induce a weakly condition which results in poorly developed fruit. They attack the thin-leaved varieties in preference to those with more robust foliage, such as the Concord, and vines of that class. Occasionally they cause much annoyance by attacking the exotic vine under glass. The Leaf-hopper seems to be more abundant at the East than at the West, and in some seasons is very numerous in the vineyards of Western New York. It has been suggested to destroy the young insect by fumigating with tobacco smoke, using a movable tent to cover the trellis and confine the smoke. When the insect can fly, it may be destroyed by carrying lighted torches through the vine-yard, though at this time most of the mischief has been done.

THE CRANBERRY.

Several insects are injurious to the Cranberry, but as these are treated of in full in the standard works on the culture of this fruit, and as they are of interest only to a comparatively small number of persons, a brief enumeration is all that need be given here. The conditions under which Cranberry culture only can be successful—the ability to flood the plantation with water, and to draw it off at will—are those which afford a remedy against nearly all of these insects. Flooding at the right time will allow the cultivator to destroy the insects that attack the vines, as well as those that injure the fruit.

THE VINE-WORM is the larva of a moth (*Anchylopera vacciniana*) which feeds upon the foliage. In Massachusetts, it hatches about the 20th of May, from eggs which have remained on the vine during the winter, and again, about the 4th of July, a second crop appears from eggs laid in June. The eggs are a flat, circular scale, of a honey-yellow color, and are deposited on the underside of the leaves.

THE FRUIT-WORM is also the larva of a moth, but a distinct and not identified species. It is of a yellowish-green color, and enters berry after berry, eating the inside of each, and finally goes into the ground to spin its cocoon, and change to a chrysalis state; unlike the Vine-worm, which spins its cocoon among the leaves at the end of the vine, drawing two together for this purpose.

The leaves are also attacked by the larva of a Saw-fly (*Pristiphora identidem*), but this insect is not numerous. The Fly makes a slit in the leaves, depositing an egg within. Broods of this species appear in June and August.

THE BUD-WORM, a small reddish-brown beetle (*Anthonomus suturalis*), about the middle of July, selects blossoms just before they are ready to expand, and deposits in them an egg through a hole made in the center of the bud. The beetle usually cuts off the bud after depositing its egg. A dull-white grub hatches from the egg, and feeds within the bud, changing to a pupa, and then to a perfect beetle, and eats its way out, leaving a round hole in the side of the bud. The beetles sometimes, though seldom, feed upon the berry. The larvæ are often killed by a minute *chalcis* fly.

Some other insects are occasionally injurious; if not disastrously so, they serve to weaken the vines and interfere with their productiveness. Among these is a Leaf-hopper (*Clastoptera proteus*, Fitch.) In its larval state, it covers itself with froth; the perfect insect jumps with the agility of a flea. Also a small Gall-gnat, the maggot of which is in some places called the "Tip-worm," as it draws together the small leaves at the tips of the growing shoots.

Insects of the Flower Garden and Green-House.

Flowering plants, whether in the green-house or in the dwelling, are subject to the attacks of several insects, which, unless they are kept in subjection, soon cause the plants to assume an unhealthy appearance. Most of the insects that infest the plants when indoors, as a general thing, remain upon them when they are placed outside during warm weather, and some of them attack hardy plants also.

In green-houses, where water can be freely used to shower the plants, and where the house can be filled with tobacco smoke as often as may be necessary, there is little difficulty in keeping the plants in a healthy condition so far as insects are concerned. Those who cultivate window plants find it more difficult to keep them free from insects by these means. Where syringing is necessary, the pots may be set in a bath-tub or sink, or, if it is desired to wet the underside of the leaves, laid upon the side, and water applied by means of a syringe, or by the use of a watering-pot with a fine rose; this should be held high above the plants in order that the water may fall with force against the foliage. All smooth-leaved plants, such as Camellias, Ivy, etc., should have the leaves occasionally washed on both sides, by the use of a sponge or soft cloth; this will not only remove the dust, but be of great service in keeping the insects in check.

House plants may be fumigated by having a large box, in which they may be shut up, and the smoke made by damp tobacco stems or other cheap form of tobacco upon a few live coals placed in an iron vessel or an old flower pot. As the use of smoke in the small way is inconvenient, and as there is a risk of injuring the plants by over-heating, it is better to apply tobacco in the liquid form. The cheapest kind of tobacco are the "stems," really the mid-ribs of the leaves, removed by the cigar makers. Either these or cheap tobacco of any other kind, may be placed in any convenient vessel and covered with water. The infusion thus made will be too strong to apply to the plants, and when used should be diluted with water until it is of the color of ordinary tea. The plants may be syringed with this, or it may be applied with the watering-pot, as suggested for the use of water. The most thorough method of using tobacco-water, and on the whole the most convenient, is to have it properly diluted in a deep tub or barrel, and to dip the plants in it, moving them up and down a few times before removing them. If this can be done once a week the plants will be kept free from most insects.

The insects which attack flowering plants in the open air only, are chiefly the Rose-bug and the Rose-slug, though grasshoppers, when abundant, are sometimes troublesome. The Rose-bug by no means confines itself to the plant from which it takes its name; it is described under the Insects Injurious to the Grape-vine on page 245.

THE ROSE-SLUG.

(Selandria rosæ, Harris.)

The main points in the history of this well-known garden pest are given by Harris in his "Insects Injurious to Vegetation," etc. It undoubtedly originated in New England, probably upon *Rosa lucida* or *R. blanda,* as these are the species of wild Rose upon which it preferably feeds. Dr. Harris first observed it in the gardens of Cambridge, Mass., in 1831, and observes that six or seven years elapsed before it made its appearance in Milton, where he then resided. It feeds only at night, except in very cloudy weather, and exclusively upon the upper surface of the leaf, from which it gnaws the soft portion, leaving the veins intact. During the day it rests motionless on the underside of the leaf.

The larval life of this insect extends over a period of fourteen days, during which it moults four times. The full-grown slug is rather more than one-third of an inch in length, by one-ninth in diameter. The thoracic joints are somewhat smaller and humped, but not puffed out laterally, as in some closely allied species, nor has it, like these, a slimy surface. The color is a translucent dull-yellow, becoming more opaque at the last moult. Soon after this it enters the ground, and incloses itself in a fragile, earthen cocoon, within which it remains dormant for many months, not changing to pupa until the following spring. Harris's assertion that it is double-brooded has long been doubted by careful observers, and is unquestionably disproved by Miss Murtfeldt's experiments.

Owing to the longevity of the flies and the different dates at which they emerge, there is a succession of larvæ, covering a period of from four to six weeks; but they are all of the same brood, and when once they have entered the ground, that is the end of them for the season.

The Rose-slug, like most other insects, has a large number of natural enemies, but these are not yet adequate to the task of keeping it in check. The attention of florists has, therefore, been largely directed to the discovery of some reliable artificial remedy.

Various applications have been tried with more or less success, among which the most certain in its effects is whale-oil soap suds, made in the proportions of one pound of soap to eight gallons of water. The objections to this remedy are, that it has a disagreeable odor and is liable to discolor the opening buds. Dusting freely with White Hellebore has also been tried with very good success, and it may be used in water, as directed for Currant worms, p. 204. The Pyrethrum powders have as yet been used only to a limited extent, but with the prospect that throughly applied they would prove effectual.

PLANT-LICE—APHIDES.

There are a great many species of plant-lice or aphides. Almost every plant is liable to the attacks of some species peculiar to itself. They are found upon the roots as well as upon the stems and leaves, where they insert their long tubular beaks and suck the juices of the plants, and only change their places when they have exhausted the sap in that locality. It would be impossible to even mention the various species in a work like this, much less to give a detailed description of them. Every farmer and gardener will know from the curled appearance of the leaves of various trees and herbaceous plants the author of the mischief.

Numerous parasites keep these destructive plant-lice greatly in check, and it is always well to look closely, before making an application to destroy the lice, to see if there are not some parasites at work, and if so they will often clear the plants much more effectively than any remedy we can apply. This I have observed both at the North and South, and usually when I have been studying other insects.

In Florida I was studying a large black and red ant (*Campanotus esuriens*), and was greatly interested in their immense droves of dark-colored aphides—the "ant's cows" as they are often called, that were thickly clustered on the underside of the young leaves of an orange tree. While watching the ants moving about among the droves, I noticed several tiny Ichneumon flies mounting the backs of the plant-lice. They were so small as to be scarcely visible to the naked eye, but a good lens soon helped me to see what they were doing. They were busy depositing eggs in the "ant's cows!" The Ichneumon would mount the back of a "cow," when the latter would become restive and try to dismount its rider by kicking and nearly standing on its head, and this would set the others next it to kicking in the same way, until all on the leaf seemed to be panic-stricken, and were kicking, striking, and throwing themselves about in a most ludicrous manner, all the while holding on by their beaks. And it was very amusing to see the excited ants trying to find the cause of the panic. But the little Ichneumons did not seem to be in the least disconcerted and did their work most effectually as the sequel proved.

Not many days after I witnessed the egg-laying, the abdomens of the plant-lice were very much distended, and they no longer gave any nourishment to the ants, who passed around among them as if discouraged. Two ants would meet and seem to consult over the matter, then they would stroke the "cows" with their antennæ, but meeting with no response they would pass to another leaf, with no better result. At last they tried to remove the "cows," they would take them gently in their mandibles, but in many cases the beak was inserted so firmly in the leaf or twig they could not remove it. When they did succeed in removing one they invariably carried it to the nest.

This was the most complete destruction of plant-lice I ever witnessed. I could not find a single living specimen left. In due time a little shining black Ichneumon-fly—the counterpart of its mother—emerged from a hole in the back of each aphis.

Since my observations were made on this orange aphis it has been named by Mr. Ashmead, *Siphonophora citrifolii*, and the little Ichneumon has been named by Mr.

Cresson as a species of *Trioxys*. Aphides, wherever they occur, are readily destroyed by the use of tobacco, applied as smoke or in infusion as already described. What is known to gardeners as the "Blue Louse" is an aphis which sometimes attacks the roots of verbenas, asters, and other flowers in such numbers as to cause their death before the source of the trouble is suspected. When these underground lice attack the roots, a persistent application of tobacco-water will save the plants if it is used before the injury has gone too far.

THE MEALY-BUG.

Genus Dactylopius.

This insect is a common pest of the green-house both in this country and in Europe, and is often injurious to plants in the open air. There are several species, all of which are more or less covered with a quantity of floury matter secreted through pores scattered over the body. They are often very abundant upon almost

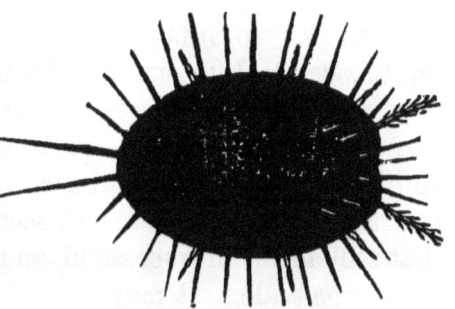

Fig. 157.—MEALY-BUG.

every variety of house-plant and very destructive. They are most frequently found in the crotches of the branches, and close down in the axils of the leaves, though they do not confine themselves to these places. The engraving, figure 157, shows a Mealy-bug, with its powdery covering removed and much magnified. One species—*D. destructor*, Comstock—is one of the worst enemies to the orange groves in Florida.

Professor Comstock, in his Report as Entomologist of

the Department of Agriculture, says: "the natural enemies of the Mealy-bug—*D. destructor*—is a little chalcis fly (*Encyrtus inquisitor*, Howard)," also "a small red bug was observed by myself and several of our correspondents to prey upon the Mealy-bug. The very curious larvæ of a lady-bird beetle, known as *Scymnus bioculatus*, were found feeding upon the eggs of the Mealy-bug at Orange Lake. These larvæ mimic the Mealy-bug so closely they might easily be taken for them."

The great difficulty in the way of destroying this insect is the floury secretion with which it is covered, most washes having little effect upon it. The best remedies, so far as I know, are given by Professor Comstock in the Report above mentioned.

REMEDIES.

"SNUFF AND SULPHUR.—Equal parts by bulk of smoking tobacco and flowers of sulphur were ground together in a mortar until thoroughly mixed. This compound was perfectly successful when dusted over wet plants; and it adhered to the plant for a long time notwithstanding rain. Still this does not seem to me to be a remedy that will admit of successful and economical application on a large scale. It may be useful in conservatories, and upon ornamental plants."

A decoction of tobacco is also useful in destroying the Mealy-bug. The Mealy-bug upon window plants and upon those in green-houses, if taken in time and perseveringly followed, may be kept in check by a modified hand-picking, removing the insects wherever they may be found by means of a small stick, such as a sliver of pine sharpened to a point. An "exterminator" is offered, but as its composition is kept secret, it can not be intelligently commended.

THE ROCKY MOUNTAIN LOCUST.

(Caloptenus spretus, Thomas.)

Though the ravages of this insect are confined to a limited area, its destructiveness is so great in the localities it visits, that it seems desirable in a work like the present to give the leading facts in its history. It is usually called the Rocky Mountain Locust, but is sometimes known as the "Hateful Grasshopper." This insect has visited Kansas, Nebraska and other Western States with most destructive effect, the recital of which reminds one of the accounts of the plagues of Egypt. Few insects have had their life history more thoroughly studied, and the useful information given by entomologists concerning this single insect has more than warranted the cost of the various State and General Government Commissions. An elaborate account of this insect is given in the Seventh Missouri Report, and another, in the Report of the U. S. Entomological Commission for 1880. The following is compiled from an account in the "American Entomologist," by Wm. A. Byers, and from other sources. The Rocky Mountain Locust is common in all the western or rainless region, one-third of the United States, but its breeding place is upon the hot, parched plains and table lands, from four to six thousand feet above the sea. The greater the heat, the more they flourish. Though they endure considerable cold and live, they are at the same time exceedingly sensitive to its effects; becoming torpid in frosty nights or in snow storms, and reviving to active life in the succeeding sunshine. The swarms that devastate the country in their flights are invariably natives of sandy plains or basins, comparatively destitute of vegetation, where the direct and reflected heat of the sun's rays in summer are more intense than are experienced in the Valley of the Mississippi. The humidity, however, is very much less; the air being like that of a furnace. In such places, and on the hottest days, the Grasshopper is the most active, and then it attains its greatest perfection. When it has reached a certain stage in its existence, it takes to flight. Those hatched in the same locality, and necessarily under the same climatic influences, rise in the air about the same time, but they do not move in concert. Their course is directed by the prevailing winds more than by any other influence. Consequently, in this country, it is generally from northwest to southeast. They alight or move forward at pleasure, each individual upon its own account. Many of them fly at an immense height. . They have been seen on the highest peaks of the snowy range, fourteen to fifteen thousand feet above the sea, filling the air as much higher as they could be distinguished with a good field glass, glistening in the sunlight like snowflakes. In crossing the snowy ranges countless myriads of them perish. Nearly all that alight for food become so chilled that they are unable to rise again, and in a few days they die. On the great snow fields it is nothing uncommon to see the dead so plentiful that they might be shovelled up by wagon loads. When the season comes for depositing their eggs, the swarms which happen to be in favorable localities, proceed to do so, after which most of them soon die and the pest disappears. Some doubtless continue their flight. If the succeeding winter is mild, young Grasshoppers may be found upon sandy, sunny hillsides long before spring, but the great swarms appear with the earliest vegetation. Then it is they are the most destructive. It is a common belief that a young Grasshopper eats more than half a dozen full grown ones. They feed and grow, and in due time take flight, as did the generation before them. But few Grasshoppers are hatched in the mountains, properly speaking. It is true they do in some of the valleys, up to the altitude of seven or eight thousand feet—possibly sometimes to nine thousand—but they usually come out so late that the frosts of the following fall catch them before they take flight. As an illustration, the Middle Park of the Rocky Mountains is a great basin, bowl-shaped, with a single line broken out of its western rim. Otherwise, it is surrounded by snowy mountains. Fifteen years ago, it was invaded by Grasshoppers from the direction of Utah, which deposited their eggs all over it. In its lower portion the young began hatching about the first of July. They attained maturity and took flight in August. Their hatching ground was from six to seven and a half thousand feet above the sea. Further up toward the rim they came out later, and at nine thousand feet they did not appear until the last of August. September frosts and snows caught them, and they never left their native ground. About the same time these latter hatched, immense swarms of full-grown insects came again from the west, but instead of lighting in the Park they drifted up against and upon the snowy range east of it, where they perished in countless millions.

In August, 1864, this country had its worst visitation of "Hateful Grasshoppers." They had hatched in the valleys of the Upper Missouri, from six hundred to eight hundred miles distant, and swept over Colorado with a solid front. They ate up late crops and then deposited their eggs and died. In the following spring, their progeny came out of the ground with the early crops, which they devoured. When about one-third grown they were attacked by an Ichneumon Fly, which stung them in the back, depositing one or more eggs. The product of these destroyed probably one-half or two-thirds of the Grasshoppers, and the balance in due time took flight and left us. With the exception of those two years, Colorado has not been generally nor severely scourged by that pest.

They have done damage in several restricted localities, and have passed over in greater or less swarms almost every year since the settlement of the country, but the prevalent idea that they are a yearly plague is a mistake. In New Mexico, which has been settled by the same people for two hundred years, generation after generation of the same family, cultivating the same fields, they say they expect to lose about one crop in seven by Grasshoppers. The experience in Utah, Montana, Idaho and

Fig. 158.—THE FEMALE ROCKY MOUNTAIN LOCUST DEPOSITING HER EGGS.

a, a, a, Female Locusts in different positions, ovipositing; *b*, Egg-pod extracted from ground, with the end broken open; *c*, Eggs; *d, e*, Earth partially removed, to show an egg-mass already in place, and one being placed; *f*, shows where such a mass has been covered up.

Nevada, is about the same as Kansas and Nebraska, which States have suffered more or less until recently. They will not propagate in great numbers in the Mississippi Valley—not because it is too hot or too low, but because it is too damp.

When the Grasshopper invades a district, it at once sets about depositing its eggs, and the great injury to be apprehended is, from the brood to be hatched from them.

EGG-LAYING AND HATCHING.—Figure 158 illustrates the manner in which the female lays her eggs. With two pair of horny valves at the tip of the abdomen, she is able to drill a cylindrical hole in the ground, preferring for this purpose soil that is rather firm, though not too hard. In a moist climate, or where vegetation is rank, she chooses bare and exposed places, but in her native range, viz., the Northwestern Plains, where the vegetation is usually scant and short, she chooses rather the shade at the base of some Sage bush or Grease-wood shrub. When the hole is once drilled the eggs are laid in four tolerably

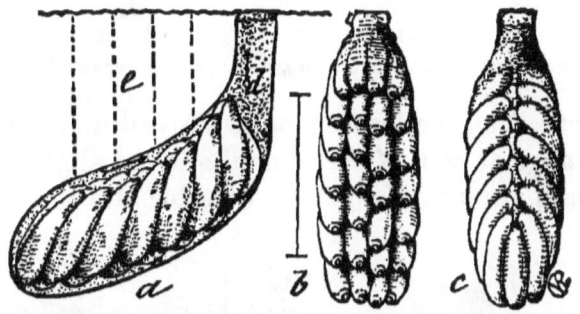

Fig. 159.—EGG-MASSES OF LOCUST, MAGNIFIED.
EGG-MASS.—*a*, from the side, within burrow; *b*, from beneath; *c*, from above.

regular rows (fig. 159), interspersed by a fluid which is frothy and mucous, and which dries around the eggs and fills up the neck of the burrow (fig. 159, *d*). Each female lays from two to three batches of eggs, each batch containing about thirty eggs. The eggs are laid throughout the late summer and fall months until winter sets in, at which time every stage of embryonic development can be found. The great bulk of the eggs remain unhatched until the ensuing spring.

HABITS AND DEVELOPMENT.—The young locusts congregate in large numbers in warm and sunny places. At night, or during cold and damp weather, they usually huddle together under any shelter or rubbish that may be at hand. They do not migrate until they have eaten off the vegetation where they hatch. This usually happens when they are about one-third or one-half grown. They then travel during the warmer hours of the day by alternately walking and hopping in vast bodies in some given direction. In thus travelling they move at the average rate of about three yards a minute. There are six stages

Fig. 160.—THE LARVÆ AND PUPA OF LOCUST.
a, a, Newly-hatched Larvæ; *b*, Full-grown Larva, *c*, Pupa of the Locust.

of growth, *i. e.*, the locust moults at five different periods. The change at each of these moults is but slight, and the wing-pads are first distinctly noticeable and turned up in

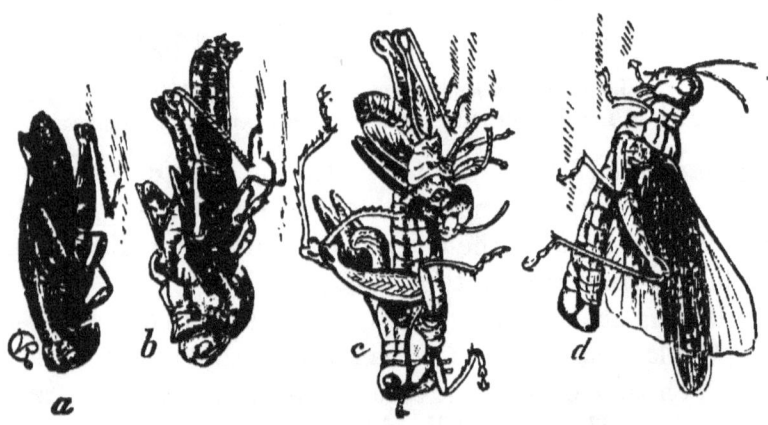

Fig. 161.—THE PUPA OF THE LOCUST ACQUIRING WINGS.
a, Pupa with skin just split on the back; *b*, the imago extruding; *c*, the imago nearly out; *d*, the imago with wings expanded.

the fourth stage, or after the third moult. After the fourth moult we have the true pupa stage (fig. 160, *c*), and with the fifth moult the wings are acquired, the process being illustrated at figure 161. The time required from hatching to full development varies according to season and weather, cold and wet weather retarding, and warm weather accelerating development. It averages, however, two months. There is but one generation each year, the term of the insect's life being bounded by the spring and autumn frosts.

Of the various methods of combating the attacks of this Grasshopper, we have

THE DESTRUCTION OF THE EGGS.—Harrowing in the autumn, or during dry, mild weather in early winter, will prove one of the most effectual modes of destroying the eggs and preventing future injury, wherever it is available. A revolving harrow or a cultivator will do excellent service in this way, not only in the field, but along roadways and other bare and uncultivated places. The object should be, not to stir deeply but to scarify and pulverize as much as possible the soil to about the depth of an inch.

PLOWING.—Next to harrowing this is one of the most generally available means possessed by the farmer of dealing with locust-eggs.

IRRIGATION.—This is feasible in much of the country subject to locust ravages, especially in the mountain fields or gardens.

COAL-OIL.—The use of coal-oil and coal-tar may be considered, as both substances are employed in various ways for trapping and destroying the insects. Coal-oil is the very best and cheapest that can be used against the locusts. It may be used in any of its cruder forms, and various contrivances have been employed to facilitate its practical operation. The main idea embodied in these contrivances is that of a shallow receptacle of any convenient size (varying from about three feet square to about eight or ten by two or three feet), provided with high back and sides, either mounted on wheels or runners, or carried (by means of suitable handles or supporting rods) by hand. If the "pan" is larger than, say, three feet square, it is provided with transverse positions which serve to prevent any slopping of the contents (in case water and oil are used), when the device is subjected to any sudden irregular motion, such as tipping, or in case of a wheeled pan, when it passes over uneven ground. The wheeled pan is used like a wheelbarrow; the hand-worked pan is carried by long handles at its ends. On pushing or carrying, as the case may be, these pans, supplied with oil, over the infested fields, and man-

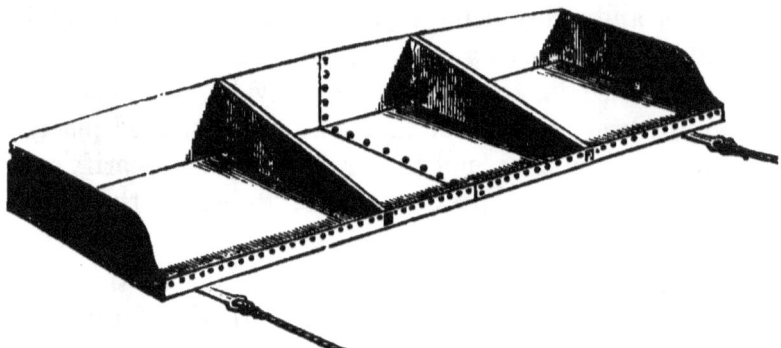

Fig. 162.—COAL-OIL PAN FOR CATCHING LOCUSTS.

ipulating the shafts and handles so as to elevate or depress the front edge of the pan as may be desired, the locusts are startled from their places and spring into the tar or oil, when they are either entangled by the tar and die slowly, or, coming in contact with the more active portion of the oil expire almost immediately. Fig. 162 represents a sheet-iron pan that has been used in some localities with good results. It must be made sufficiently tight to hold kerosene, of which sufficient is used to cover the bottom. A simpler form of pan is shown in figure 163. The bottom of this is to be covered with a thin layer of coal tar. Pans of this kind are made light enough to be drawn across the fields by boys; or if heavy, horses are used to drag them. The majority of the insects perish within the pans, which must be occasionally emptied. If some of the locusts jump out, it is of little consequence, as all that have been touched by the oil will soon die. In Colorado they use kerosene to good advantage

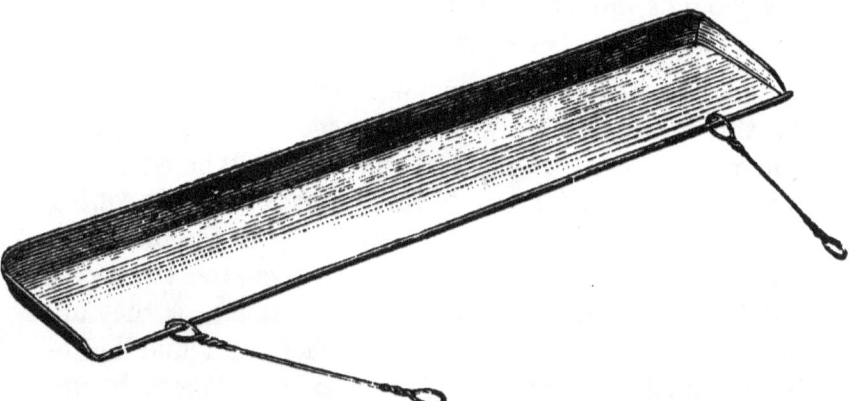

Fig. 163.—COAL-TAR PAN FOR CATCHING LOCUSTS.

on the water in their irrigating-ditches, and it may be used anywhere in pans or on cloths, stretched on frames and saturated with it, to be drawn over the field.

DESTRUCTION OF THE YOUNG OR UNFLEDGED LOCUSTS.

1st. BURNING.—This method is perhaps the best in prairie and wheat-growing regions, which compose the larger part of the area subject to devastation by this locust. In such regions there is usually more or less old straw or hay that may be scattered over or around the field in heaps and windrows, and into which the locusts, for some time after they hatch, may be driven and burned. During cold or damp weather they congregate of their own accord under such shelter, when they may be destroyed by burning, without the necessity of previous driving. Much has been said for and against the beneficial results of burning the prairies in the spring. This is chiefly beneficial around cultivated fields or along the road sides, from which the locusts may be driven, or from which they will of themselves pass for the shelter the prairie affords.

As locusts disperse more and more from their hatching-grounds into the prairie as they develop, burning the grass in spring is beneficial in proportion as it is delayed.

2nd. CRUSHING.—The wholesale destruction of locusts by this means, can only be advantageously accomplished where the ground is smooth and hard. Where the surface of the ground presents this character, heavy rolling can be successfully employed, especially in the mornings and evenings of the first eight or ten days after the newly hatched young have made their appearance, as they are generally sluggish during these times, and huddle together until after sunrise. It is also advantageously employed during cold weather at any time of day, since the young when the temperature is low seek shelter under clods, etc. Various machines have been devised for crushing the young.

3rd. TRAPPING.—This can easily be accomplished, especially when the locusts are making their way from roads and hedges. The use of nets at sunrise, or long strips of muslin, calico, or similar materials, converging after the manner of quail-nets have proved very satisfactory. By digging pits or holes three or four feet deep, and then staking the two wings so that they converge toward them, large numbers may be secured in this way after the dew is off the ground, or they may be headed off when marching in a given direction. Much good may be accomplished by changing the position of the trap while the locusts are yet small and congregate in isolated or particular patches.

DITCHING and TRENCHING properly come under this head; and both plans are very effectual in protecting crops against the inroads of travelling schools of the insects. They were found especially advantageous in much of the ravaged country in a year when there was little or no hay or straw to burn. They are the best available means when the crops are advanced, and when most of the other destructive methods so advisable early in the season can no longer be effectually used. Simple ditches, two feet wide and two feet deep, with perpendicular sides, offer effectual barriers to the young insects. They must, however, be kept in order so that the sides next the fields to be protected are not allowed to wash out or become too hard. They may be kept friable by a brush or rake.

The young locusts tumble into such a ditch and accumulate and die at the bottom in large quantities. In a few days the stench becomes great, and necessitates the covering up of the mass. In order to keep the main ditch open, therefore, it is best to dig pits or deeper side ditches at short intervals, in which the locusts will accumulate and may be buried. If a trench is made around a field about hatching-time, but few locusts will get into that field until they acquire wings, and by that time the principal danger is over, and the insects are fast disappearing. If any should hatch within the inclosure, they are easily driven into the ditches dug in different parts of the field.

PROTECTION BY BARRIERS.—Where ditches are not easily made, and where lumber is plentiful, a board fence two feet high and with a three-inch batten nailed to the top or side from which the locusts are coming, the edge of it smeared with coal-tar, serves as an effectual barrier, and proves useful to protect regions, where, save in exceptionally favorable locations, agriculture can be successfully carried on only by its aid, and where means are already extensively provided for the artificial irrigation of large areas. Where the ground is light and porous, prolonged and excessive moisture will cause most of the eggs to perish, and irrigation in autumn or in spring may prove beneficial.

4th. TRAMPING.—In pastures or in fields where hogs, cattle, or horses can be confined when the ground is not frozen, many if not most of the locust-eggs will be destroyed by the rooting and tramping.

5th. COLLECTING THE EGGS.—The eggs are frequently placed where none of the above means for destroying them can be employed. In such cases they should be collected and destroyed by the inhabitants, and the State should offer some inducement in the way of bounty for such collection and destruction. Every bushel of eggs destroyed is equivalent to a hundred acres of corn saved, and when we consider the amount of destruction caused by the young, and that the ground is often known to be filled with eggs; that, in other words, the earth is sown with seeds of future destruction, it is surprising that more legislation has not been had, looking to their extermination.

One of the most rapid ways of collecting the eggs, especially where they are numerous and in light soils, is to slice off about an inch of the soil by trowel or spade, and then cart the egg-laden earth to some sheltered place where it may be allowed to dry, when it may be sifted so as to separate the eggs and egg-masses from the earth. The eggs thus collected may easily be destroyed by burying them in deep pits, providing the ground be packed hard on the surface.

THE PROTECTION OF FRUIT TREES.

The best means of protecting fruit and shade trees deserves separate consideration. Where the trunks are smooth and perpendicular they may be protected by whitewashing. The lime crumbles under the feet of the insects as they attempt to climb, and prevents their getting up. By their persistent efforts, however, they gradually wear off the lime and reach a higher point each day, so that the whitewashing must be often repeated. Trees with short, rough trunks, or which lean over, are not very well protected in this way. A strip of smooth, bright tin answers better for the same purpose. A strip three or four inches wide brought around and tacked to a smooth tree will protect it, while on rougher trees a piece of old rope may first be fastened around the tree with small nails, and the tin tacked to the rope, so as to leave a portion of it both above and below. Passages between the tin and the rope or the rope and tree can then be blocked by filling the upper area between the tin and tree with earth. The tin must be high enough from the ground to prevent the 'hoppers from jumping from the latter beyond it, and the trunk below the tree, where the insects collect, should be covered with some coal-tar or poisonous substances to prevent girdling. This is more especially necessary with small trees, and coal-tar will answer as a preventive.

One of the cheapest and simplest modes is to encircle the tree with cotton batting, in which the insects will entangle their feet and thus be more or less obstructed. Strips of paper covered with tar; stiff paper tied on so as to slope roof-fashion; strips of glazed wall-paper, and thick coatings of soft-soap, have been used with varying success; but no estoppel equals the bright tin. The others require constant watching and renewal, and in all cases coming under our observation some insects would get into the trees, so as to require the daily shaking of these morning and evening. This will sometimes have to be done when the bulk of the insects have become fledged, even when tin is used, for a certain proportion of the insects will then fly into the trees. They do most damage during the night, and care should be had that the trees be unloaded of their voracious freight just before dark.

It has been found that the whitewash was rendered still more effectual by adding one-half pint of turpentine to the pailful.

DESTRUCTION OF THE WINGED INSECTS.

The complete destruction of the winged insects, when they swoop down upon a country in prodigious swarms, is impossible. Man is powerless before the mighty host. Special plants, or small tracts of vegetation may be saved by perseveringly driving the insects off, or keeping them off by means of smudges, as the locusts avoid smoke; or by rattling or tingling noises constantly kept up. Long ropes perseveringly dragged over a grain field have been used to good advantage.